Project UNIVERSE

Project UNIVERSE

An Experiment in High-speed Computer Networking

J. W. Burren
and
C. S. Cooper
Advanced Communications Unit,
Rutherford Appleton Laboratory

CLARENDON PRESS · OXFORD
1989

Oxford University Press, Walton Street, Oxford OX2 6DP
Oxford New York Toronto
Delhi Bombay Calcutta Madras Karachi
Petaling Jaya Singapore Hong Kong Tokyo
Nairobi Dar es Salaam Cape Town
Melbourne Auckland
and associated companies in
Berlin Ibadan

Oxford is a trademark of Oxford University Press

Published in the United States
by Oxford University Press, New York

British Library Cataloguing in Publication Data
Burren J. W.
Project UNIVERSE: an introduction to
high-speed computer networking
1. Computer systems. Communication
networks
I. Title II. Cooper, C. S.
004.6
ISBN 0-19-853723-9

Library of Congress Cataloging in Publication Data
Burren, J. W.
Project UNIVERSE: an introduction to high-speed computer
networking, J. W. Burren and C. S. Cooper.
p. cm. Bibliography: p.
Includes index.
1. Computer networks. 2. Project UNIVERSE. I. Cooper C. S.
II. Title.
TK5105.5.B87 1989 004.6 – dc 19 88-29048
ISBN 0-19-853723-9

Printed in Great Britain
by Biddles Ltd,
Guildford & King's Lynn

Foreword

UNIVERSE, the project described in this book, was in many respects unusual and in some unique. It was a very substantial systems experiment in computing and communication—at £5M perhaps not vast by the standards of pure science but definitely so as engineering—and was a real experiment rather than an attempt to make a service system. It showed, among many other things, that such experimentation is feasible; it has to be admitted, though, that a certain amount was lost by not being able to give a real service to a community. UNIVERSE was also a large collaboration, undertaken before major collaborations became fashionable in Information Technology, and furnishing some of the experience on which later collaborative programmes were based. In the view of the UNIVERSE partners, who later took part in other collaborative programmes, changes to the administrative mechanics were far from being uniformly the improvements they were presumably intended as. Another aspect demonstrated by UNIVERSE was the immense importance of the pressure brought on decision-makers by the project's dependence on a time-limited resource—the OTS satellite. It was made very clear to the numerous bodies that had to approve the project that any delay in decision was equivalent to rejection; this had an extremely beneficial effect on their deliberations.

Once the project had started, the time-pressure was a great help in keeping up momentum. People worked very hard and there was a constant to-and-fro of tasks; it didn't matter who did what as long as the job was done in the shortest time. The experimental programme ran up to the last minute the satellite was available, some unspent balances being used to buy extra satellite access in the last couple of weeks. When one looks today at the experiments done, they have a modern look despite being over four years old. This is because the UNIVERSE vision of exploiting wide-area networks as if they were local was very timely, and in discussions of research-level distributed computing today, one can often make contributions by recalling how something was done in UNIVERSE. It would be inappropriate to go into detail here, but one may perhaps instance the study of cooperation between authentication services at different sites with varying levels of mutual trust—quite a hot topic of later years. The UNIVERSE internetwork architecture, too, has been invaluable as a real and working example of something notably different from conventional approaches. Emphasis on multimedia use of digital networks has a thoroughly modern ring today.

It is aspects such as these that make it worthwhile producing a book on UNIVERSE rather than leaving the record to some forty publications in the literature.

Cambridge *R. M. Needham*
August 1988

Preface

This book describes a collaborative endeavour entitled 'UNIVersities Expanded Ring and Satellite Experiment' which took place in England during the years 1981 to 1983. The collaborators were drawn from universities, industry, and a government laboratory. Funding was provided by the Science and Engineering Research Council, the Department of Trade and Industry, British Telecom, GEC Ltd, and Logica Ltd.

The idea for the project arose from experience that one of the participants had gained with satellite transmission. It was observed that apart from the transmission delay, there were remarkable similarities between a broadcast satellite network with forward error correction and local area networks. This gave rise to the thought that perhaps it might be possible to combine the technologies to provide a network demonstrating many of the properties of a local area network but on a continental scale.

This idea was conceived in 1980. The work reported here was essentially completed during the years 1981 to 1983, although there had been some preparatory technical work and a great deal of fund raising activity during 1980. There were some final pieces of detail investigated in 1984 and a great deal of report writing took place during that year but in essence the experiment ended on 31st December 1983 when the Orbital Test Satellite was taken out of service.

It is our intention that this book should serve both as a record of the technical aspects of the project and as a case study for this type of collaborative endeavour. While we have been at some pains to try to present reasonably balanced coverage of all aspects of the project, ultimately the view presented is the personal one of the authors. Obviously, such an endeavour could not have been undertaken without a great deal of co-operation among all those participating. In retrospect, a lasting impression left by the project was the degree of single mindedness with which the project was undertaken. This, together with the sense of urgency induced by the ever looming prospect of the satellite channel being 'switched off', made the project an exciting one to be involved in.

Throughout this book reference is made to a number of different areas of computer and communications science and technology, all of which are components of the formative field of study now generally referred to as information technology. Generally, we have tried to include sufficient background material to make the book reasonably self-contained, supplementing this where necessary by reference to review articles and established texts in the particular subject areas. The second edition of the book by Tanenbaum (1988) may be mentioned in particular as an excellent general text covering most aspects of computer networking.

During the final months of preparation of this book, two other works of particular relevance came to our attention. One of the concepts at the heart of the project was that of enlarging the distance over which distributed systems principles could be made to operate. The first work, Coulouris and Dollimore (1988), provides an excellent introduction to the distributed systems principles assumed implicitly by us; it also provides many examples and case studies from a number of research systems in distributed computing.

The second work, Comer (1988), ably fulfils a long-standing need for an introduction to the architecture, ideas, and TCP/IP suite of protocols used in the Internet developed by the Defense Advanced Projects Agency (DARPA) of the Department of Defense in the USA. Clearly, no project such as UNIVERSE can take place without building on the wealth of knowledge and experience already contributed by others. If ideas on distributed systems could be said to form one cornerstone of the project, so packet-based computer networking is another. A rich source of ideas, results, and practical experience in this field stems from the DARPA programme of packet networking research of some twenty years' standing. The more tangible results of this programme are the DARPA Internet and the TCP/IP protocols. Although undoubtedly of great importance in forming the background against which UNIVERSE was conceived, the ideas developed in UNIVERSE can be viewed in some technical respects as an alternative approach.

Acknowledgements

Firstly, it is a pleasure to extend our thanks to all our colleagues who made the project the exciting and successful venture that we consider it to have been. We should also like to thank the many people who helped to launch the project. In particular, David Thomas from the Rutherford Appleton Laboratory, together with Mike Rogers of the University of Bristol and John Taylor of the Admiralty Surface Weapons Establishment, chairmen of the relevant Science and Engineering Research Council committees at the time, were instrumental in gaining the support of the Council for the project. David Withers of British Telecom, Lawrence Clarke of GEC, and David Mann of Logica were chiefly responsible for their companies joining the project. We should like also to mention particularly the contribution made by Robin Lingard of the Department of Trade and Industry, who was instrumental not only in obtaining support from the Department but also in pioneering the arrangements for joint funding by the Science and Engineering Research Council, the Department of Trade and Industry, and industry which were enjoyed by the project and later became widely established in the UK during the subsequent Alvey programme.

In writing this book we have, of course, drawn freely on much of the material, published and unpublished, associated with the project: the use of this source material is gratefully acknowledged. In particular, we should like to thank the European Space Agency for permission to use two of their diagrams as the basis for our Figures 2.3 and 2.4 and also to base Section 4.5 substantially on Burren (1985); Gill Waters and Elsevier Science Publishers for permission to reproduce Figures 4.3, 4.4, 4.5, and 4.6 from Waters (1984); and Chris Adams, Stephen Ades, and the Institute of Electrical and Electronic Engineers for permission to reproduce Figures 8.3 and 8.4 from Adams and Ades (1985). In addition, it is a pleasure to acknowledge our use of the four UNIVERSE papers given at the SIGCOMM '84 conference in Montreal and published by the Association for Computing Machinery (Cooper, 1984; Leslie *et al.*, 1984; Waters and Adams, 1984; Waters *et al.*, 1984) as major sources for Chapters 3 and 4 and Sections 7.4 and 9.7.5 (including Figure 7.5 and Table 9.1). A bibliography of papers published by members of the project is given in Appendix D.

Oxford *J. W. B.*
December 1988 *C. S. C.*

Contents

1. Introduction

1.1 Background

The last decade or so has seen the steady development of the enabling technologies for the growth of the techniques and products now generally labelled 'Information Technology'. An especially important component of these developments is the gradual merging of a number of different sorts of communication. Historically, the form of communication longest and best established is voice, which was traditionally based entirely on analogue techniques. Moreover, within the more specific context of the telephone industry, the emphasis has been primarily on bilateral communication based on circuit switching. The circuits for this form of communication do not have to be set up particularly rapidly, and the process was at first manual and then, for many years, electro-mechanical.

The advent of digital techniques and the development of successively larger scales of circuit integration have affected this industry technically in two fundamental ways. One is the introduction of digital encoding modulation schemes, which have encouraged the application of digital techniques to an ever increasing proportion of the necessary signal processing. The other is the application of digital computing to the problem of circuit switching and the consequent development of exchanges at the heart of which is a computer.

As a distinct and parallel activity, there has taken place in the last two decades the development of mechanisms for enabling computers to exchange data over communications channels. As in the case of communication between people over the telephone network, the scenario is generally one in which there is a single physical attachment between the computer and the communications network. Also, as for people, a single computer will generally have need to communicate with a number of other systems during its operation. In contrast to human communication over the telephone network, it is quite common, indeed even usual, for a computer to be in communication simultaneously with a large number of other systems. An even greater difference is to be found in the timescales associated with such communication. In the case of human communication, the establishment of a conversation in a second or two is regarded as quick. In computer terms, this would probably be regarded as ranging from disastrous to barely acceptable. The requirements range currently from a few milliseconds to about a second. Again, there is a great difference in the length of time for which a communication path may need to be established. For human telephone communications, the duration of a call may typically last from a minute or two to, say, an hour. The corresponding duration for computers varies from hundredths of a second to essentially permanent.

There is one other very important difference. In the case of voice communication, very little variation in transmission delay can be tolerated: in practice, this means that the necessary bandwidth must be available on an essentially permanent basis for the duration of the call. The situation for computer communication is quite different. Computers communicate in bursts and, in many cases, there is quite a wide tolerance in the delay which may be accommodated before the communication becomes so degraded as to be inadequate.

The consequence of these differences has been that much of the development of computer communication has followed somewhat different lines from that of the telephone network. The most fundamental difference is to be seen in the almost universal use of packet switching in the context of computer communication, and the consequently considerable effort that has been expended in the development of the packet switching exchange.

In the context of current developments in information technology, one may pose the question of how these two strands of communication development might be merged further into a single, coherent communications system. Envisaged is a communications network, to which a single point of attachment may be made by a subscriber, sufficient for all forms of communication. Such a network system is also envisioned to be the basis of value added services which could be based on varying admixtures of computer, video, and voice communication.

It has been suggested above, in the most general terms, the way in which digital and computer processing techniques have been influencing the development of communications technology. Much more recently there has begun a complementary evolution, whereby the use of specialized forms of high-speed communication has encouraged the experimental development of so-called *distributed computing systems*. Traditionally, a computer system consists of a processor, memory, and a number of peripherals, such as disks, keyboard displays, and printing devices. This collection of hardware is under the control of some software, resident in the processor memory (and possibly disk backing storage), known as the operating system. It is the purpose of the operating system to execute a number of support tasks which together make up the total functionality of the system. Among these tasks are support for a disk-based filing system; control of output devices such as printers and plotters; provision of spooling, for holding temporary files on their way to further processing or disposal via slow output devices, such as those just mentioned; handling user input, by keyboard and 'mouse', for example, and user output display screen. Intimately connected with the means of communication with the user is the provision of the user environment: the means by which the user interacts with the system to get work done. The work may be scientific, clerical, financial, computer programming, and so on: in each case, the user will need to have available a different set of tools and, perhaps, a different form of input handling and output display. The task which provides such environments is sometimes called the user 'shell' and is also generally considered part of the operating system. These are only some of the functions which have to be provided but they will serve to illustrate the point. The provision of an operating system capable of supporting all these functions at once presents a complex problem with a number of conflicting requirements. It is necessary to keep the output peripherals busy at all times since their slowness is in any case often a bottleneck. It is also desirable to present the impression simultaneously to a number of interactive terminal users that they have immediate and total access to the whole range of facilities of the system. It may furthermore be necessary to continue processing large production jobs submitted to a background batch facility; typically, such jobs make substantial demands on the memory, processing power, and peripheral capability of the system. Clearly, a number of these requirements are in conflict, both from the point of view of competition for resources and in the solutions which may be adopted in resolving this competition.

At the end of the 1960s and during the first half of the 1970s, there took place the development of high-speed, local communication techniques which came to be called local area networks (LANs). LANs presented the attractive possibility of separating out identifiable tasks of a system and placing these in separate processors. The advantage of this was the possibility of building dedicated subsystems in each of which fewer conflicting requirements had to be satisfied. Users could really be allocated their own processors (permanently or on a temporary basis) and would not interfere so much with each other. A single subsystem would handle disk filing. Another would concentrate on the printing of documents. Yet a third would handle a plotter. And batch processing would also happen on one or more additional systems dedicated to the purpose.

During the mid-1970s the development of very large scale integration (VLSI) technology, particularly in relation to the microprocessor, reached the stage where price, availability, and performance made it practical to begin the realization of such distributed systems. The emergence of the word processor and, more recently, engineering computer-aided design and scientific workstations which share peripherals such as printers, plotters, and disks are examples of systems based on such principles. A feature that each of these systems have in common is that they have as their object the application of computers to the office work of secretaries, engineers, and scientists manipulating the information appropriate to their profession.

It is against this background that the Universities Expanded Ring and Satellite Experiment (UNIVERSE) was conceived, in the summer of 1980.

1.2 Project overview

At the heart of the project was the idea that by marrying two communication technologies, broadcast satellite and LAN, it might be possible to construct a high-speed wide area network with many of the properties generally associated with LANs. A particular claim sometimes made for LANs is that they have the potential to carry many, perhaps all, forms of communication traffic. While as yet standing in much need of substantiation, this idea is intrinsically attractive and became another of the central themes of the project.

To take these suggestions a stage further, what was proposed was to exploit the high-speed and small delay properties of LAN technology to provide a network capable of carrying, specifically, voice and slow scan video images as well as various forms of computer traffic. Across a single LAN there were already in existence many examples of the latter. These ranged from conventional full duplex byte stream protocols carrying interactive terminal sessions or file transfer, to connectionless protocols providing file-server access, name-to-address translation, and a variety of other services relevant to the realization of distributed operating systems. Given that all of these either already existed in working form or could be developed during the lifetime of the project to operate across a single local network, they would then serve as test and demonstration traffic for a system of interconnected LANs. Existing digital encoding techniques were quite sufficiently well developed, to the extent of specialized chips being available for the purpose, to make construction of an interface between a microprocessor and a telephone handset in the form of a peripheral controller a perfectly feasible proposition. Equally, a straightforward bit-map image encoding could be used to hold pictures in computer memory in digital form. Well-known techniques could be used to interface both a camera and a raster video display to such a framestore. The whole could be controlled by a processor and interfaced to a LAN.

Thus there seemed every reason to suppose that a testing variety of traffic could in fact be provided to prove the system and, as it turned out in practice, also to demonstrate publicly the operation of the network.

The choice of broadcast satellite as the medium by which to link local networks over a wide area had its origin in a number of factors. Technically, several technologies offered potentially interesting possibilities for realizing the long-haul segments of the network, the two most obvious candidates being optical fibre links and a satellite channel. Optical fibre communication has the potential for very high bandwidth; it also has high immunity to interference. In comparison, satellites offer intermediate bandwidth; they are also susceptible to both man-made interference and noise due to adverse radio propagation conditions. A characteristic feature of satellite communication is its intrinsic broadcast nature. Moreover, satellite links can be established from mobile stations, either on land or sea; they can also be used to provide communication across land masses without any other form of communications infrastructure, whether because of hostile terrain or any other reason.

The use of a wide area broadcast medium to link a set of LANs is an intrinsically interesting research topic. However, it is also probably true to say that in practice a critical factor for the UNIVERSE project was the existence of a suitable satellite already launched and operationally available, whereas no high-speed terrestrial links were then generally available, either of optical fibre or of any other sort. Indeed, it is only since project UNIVERSE was completed that high-speed terrestrial links, such as British Telecom's 2.048 Mbit/s Megastream, have become available. By contrast, at the time of the project's inception, the European Orbital Test Satellite (OTS) had already been in operation for three years and indeed one of the participating organizations (Rutherford Appleton Laboratory) had experience of using it in an earlier experiment (STELLA project, see Section 1.3).

Although the original focus of the project was concentrated upon use of the broadcast satellite to link the local networks at each site, the value of having at least one high-speed, terrestrial link in the experimental network had always been recognized. As the project evolved, this came to be realized in practice. Two of the participants, University College London and Logica Ltd, had premises in London situated less than two kilometres apart. The proximity of these sites afforded one of the other participants, British Telecom (BT), the opportunity to provide a special, dedicated, direct link between the two.

The prime motivation in constructing the UNIVERSE network was to develop and demonstrate mechanisms for joining the individual subnetworks in such a way as to extend as far as possible the essential LAN properties of the constituents to the overall network. The essential element used to form a join between two subnetworks of the UNIVERSE network is termed a *bridge*. Bridges encompass the essence of the UNIVERSE network design. In 1980, prior to the start of the project, consideration of the construction of bridges in two different contexts had begun independently. I. M. Leslie and R. M. Needham had worked on the problem of bridge design for the interconnection of Cambridge rings. (This work came to fruition during the project and is described in Leslie, 1983.) Simultaneously, the use of a broadcast satellite network for the interconnection of LANs had been considered by J. W. Burren and others at Rutherford Appleton Laboratory (Adams *et al.*, 1982). The two groups had proposed essentially identical schemes for the two types of bridge and it was the amalgamation of these schemes into a single design that formed the basis of the architecture used for the UNIVERSE network. A full discussion of both the constituent subnetworks and the bridges is the subject of subsequent chapters; here, it will suffice to draw attention to the salient features.

An obviously desirable design goal for bridges is that they should be fast. One factor in achieving this at reasonable cost is to reduce to a minimum the amount of processing which must be expended in forwarding data from one subnetwork to another. A second key consideration for bridges was that they should not impose strategies such as retransmission upon all traffic since, apart from the processing load implied, this would make their use impossible for any type of real-time traffic where low and predictable delay is of greater importance than either the data being correct to a very high probability or, indeed, the guaranteed delivery of every data packet. The final consideration characteristic of the UNIVERSE bridge design was to preserve the order of packets being transported between stations on the network. This is an intrinsic property of both LANs and the broadcast satellite channel: in each case, blocks may be lost (though this is extremely rare) but blocks transported between two given stations will arrive in the same order as they were transmitted. This last requirement is one which primarily affects routing decisions. In the UNIVERSE network the topology was in any case very simple, only one site having the possibility of a loop formed by bridges. The decision was made essentially in order to guarantee simplification of the job of hosts wishing to construct reliable, sequenced transfer of streams of data. The project explicitly chose not to focus upon routing issues. The outcome of these key design considerations was the concept of the *lightweight virtual circuit* (LVC), which will be elaborated fully in subsequent chapters. The essence however was that a fixed route was formed at the time of translating a name to an address and that neither error recovery nor flow control was provided by the bridges providing the circuit.

The lightweight virtual circuit was the principal network access protocol provided. However, several others were also made available for various experimental purposes: these were *single shot protocol* (SSP), *datagram*, and *broadcast*. Single shot provides, in its basic form, the exchange of a single packet in each direction between two hosts, forming a request and reply transaction. The importance of this particular form of interaction arises from its considerable use in distributed systems design. Obviously such an interaction can be built upon datagrams. In UNIVERSE, SSP was provided in order to support systems existing at Cambridge University Computer Laboratory (and to a lesser extent Rutherford Appleton Laboratory) which it was proposed to exercise over the wider UNIVERSE network. It is also, incidentally, relatively easy to provide, given that lightweight virtual circuits are being provided, since it is very little different from the initial setup exchange for such a circuit.

In any communication network an important consideration is the question of how to deal with congestion, in both the constituent channels and the switching elements. The techniques traditionally adopted to overcome this problem are generally specific both to the nature of the traffic and to the network. In a more general network the problem is exacerbated and there is no generally accepted solution. The specific strategy adopted in the UNIVERSE network was for the bridges to discard packets in the face of congestion. Experience with this approach will be described in later chapters.

Since by its very design the UNIVERSE network was intended to be able to carry a wide range of communication traffic, it was in particular able to carry a range of computer communication protocols. This was seen as presenting an opportunity to implement and test a variety of heterogeneous protocols to gain experience of their use. Perhaps more significantly, some initial experience was gained in the use and observation of the behaviour of a particular network when subjected to a mixture of traffic with such contrasting needs as

(a) traditional data exchange between computers, where high bandwidth and low delay are desirable, if ultimately a luxury, whereas loss or corruption of data is a disaster; and

(b) real-time, conventional voice, where loss or corruption simply degrades the service, whereas any substantial or unpredictable delay or lack of necessary bandwidth rapidly renders the communication incomprehensible.

By the time the experimental phase of the project ended, at the end of 1983 when the OTS was taken out of service, a limited amount of experience had been obtained in using the network for quite a wide variety of communications traffic.

- Conventional data interchange over reliable circuits was used for both remote terminal access to experimental systems and for file transfer. While no particular experiments were conducted to tune such circuits to take account of satellite delay, the mechanism was used quite heavily, both for remote systems support and as a means of setting up and controlling distributed experiments.

- Transactional exchange traffic was used quite widely, especially as part of the interaction of individual essential components of the network and also in the operation of a variety of distributed systems.
- Within the limits imposed by the speed of operation of the experimental UNIVERSE bridges, one of the most successful demonstrations was in using the network to transfer video stills and slow-scan video captured by remotely controlled cameras.
- A number of experiments were carried out on the ability of the network to carry voice traffic. Little difficulty was experienced in the development of a recorded-message server which could be used both locally and remotely across the satellite segment. Considerably more difficulty was both anticipated and encountered in achieving real-time conversational mode. More work would be necessary to turn the system into a service; however, useful new insight was gained into the nature of packet voice communication.
- Several of the participating organizations seized the opportunity to implement and test protocols for carrying videotex, teletex, encrypted data, and IBM 3270 protocol. In these developments, the network simply acted as a convenient experimental vehicle, while the existence of the project offered an opportunity for pre-competitive collaborative development among the commercial participants.
- One of the more novel, and incidentally entertaining, demonstrations of the network was in primitive remote robot control. While the network could certainly never be used (at least remotely over the satellite segment) for interactive, real-time control, it was nevertheless remarkably easy to use for predetermined absolute remote control without feedback. These developments were undertaken primarily to afford a means of public demonstration of the capabilities of the network. Nevertheless, the simplicity of their implementation was a consequence of the ease of use of the transactional SSP access to the network.

1.3 Participants

The initial concept of the project was developed during the summer of 1980. Since the idea originated within the academic community, it was at first seen purely as an academic research project and a proposal was put to the Science and Engineering Research Council (SERC). The proposal was a joint one by the Computer Laboratory at the University of Cambridge, the Department of Electrical and Electronic Engineering at Loughborough University of Technology, Rutherford Appleton Laboratory (RAL), and the Department of Computing Science at University College London (UCL).

The background and previous achievements of the participating groups was sufficiently complementary to be able to cover in some depth essentially all technical aspects of the proposed undertaking. At Cambridge, operating systems and the development of the Cambridge empty-slot ring had for some years been major topics of study. In the years immediately preceding the project, these two areas of work had come together and had resulted in the development of the Cambridge distributed computing system (Needham and Herbert, 1982). The specific operating system which had acted as the primary vehicle for the implementation of this distributed system was TRIPOS (Richards *et al.*, 1979), a portable, research operating system originally developed for minicomputers and subsequently transferred to microcomputers. The Department of Electrical and Electronic Engineering at Loughborough had considerable experience and a continuing research programme in the development of modems, particularly digital modems, coding schemes for speech and image transmission, and the construction of digital image stores. This expertise was a valuable complement to that of the other academic participants whose backgrounds were primarily in computer science. Members of the group at RAL had contributed to the development of the SERC computer network and to the design of the initial set of UK 'Coloured Book' protocols (PSS User Forum, 1980 and 1981, and DCPU, 1981*a* and 1981*b*) adopted for the UK Joint Academic Network (JANET) (Wells, 1984). The group had also been participating in the European Commission's Satellite Transmission Experiment Linking Laboratories (STELLA) (Hine, 1984), which had investigated the use of Ku-band satellite communication to provide a high-speed, low error rate, bulk data transmission facility. Finally, UCL had for many years been contributing to computer network research and development, both in the UK and the USA and had been responsible for the gateway linking the US Department of Defense Advanced Research Projects Agency's network, ARPAnet, to networks in Britain. UCL were major participants in the DARPA SATNET project (Kahn, 1979; Jacobs *et al.*, 1979; Jacobs *et al.*, 1978); they had also contributed to both the development of the UK 'Coloured Book' protocols, as well as to the DARPA Internet programme. The influence of many of these topics of research pursued by the individual participants can be seen in aspects of the UNIVERSE project.

During its consideration by SERC in the autumn of 1980, it was suggested that the project was a suitable one in which to involve industry directly and that the Department of Trade and Industry (DTI) should be approached for part of the funding. The principle of such a collaboration was at that time something of a new departure, though there have since been numerous examples of similar collaborative endeavours, both nationally in Britain under the auspices of the Alvey programme and in a European international context in programmes such as ESPRIT (European Strategic

Programme for R&D in Information Technology). The idea behind all such collaborations is generally to foster closer links between academia and industry. However, in the field of computing and communication technology, where developments occur with astonishing rapidity, there is the more specific goal of providing a mechanism whereby promising results of research and development may more rapidly be applied in the development of new products and services.

In the event, during 1981, three further partners joined the project, all from industry. GEC's Marconi Research Centre at Chelmsford, one of two research laboratories associated with an information systems group of companies within GEC, has a wide remit for research and development in information technology. In particular, one of the groups there had become interested in the technology of LANs and was considering the design of a token ring. This arose out of a specific interest in the application of LAN technology to manufacturing automation. Although the UNIVERSE network was to be constructed using Cambridge ring LANs, there was no intention of making the architecture dependent upon this or any other particular LAN technology. An attractive way of ensuring this was the possibility that GEC might be in a position to connect their token ring. The GEC interest in manufacturing automation brought to the project a perspective complementary to that of most of the other participants.

British Telecom Research Laboratories (BTRL) at Martlesham Heath, near Ipswich, are occupied in research and development of all kinds of communication and information systems. The particular groups which became involved in the UNIVERSE collaboration were those responsible for videotex (Prestel) and teletex development. Mounting these services on a LAN-based network could be seen as one possible way of introducing them into an office context by means of a single form of communication carrier. A further, initial impetus in the case of Prestel was the possibility that a network such as UNIVERSE, capable of broadcast, might offer a significant new approach to the problem of updating multiple copies of the same database situated all over the country.

Logica Ltd is a consultancy firm with international subsidiaries in a number of countries. Their interests have included work on various systems relating to satellite operations, office systems, and computer communications. A UK subsidiary, Logica VTS Ltd, was at the time one of the manufacturers of a commercial form of the Cambridge ring. As the programme for the project was elaborated during the first year (while the network itself was being built), Logica's interest developed in three general areas: business communication, image compression for communication, and the construction of a program development environment using a set of widely distributed resources.

1.4 Organization

Although the project was approved during the first quarter of 1981, the precise details of funding, management, and mechanisms of collaboration were not all decided and in place until much later in the year. The nature of both the funding and the collaboration was original and the process of working out how the whole endeavour should be managed took some time. Since this account of the project is intended to cover all aspects, a short description of the management and funding arrangements is given here. It is hoped that by so doing, others engaged in similar undertakings may be enabled to build upon the project's experience. The purposes of the various committees and working groups are outlined here; they are listed in Appendix A.

The funding for the project was eventually agreed on the following basis. The total cost of the project was, in round terms, about £5M. The DTI, which was at the time building up to support a number of programmes and initiatives in information technology, agreed to fund 50 per cent of the project. The sole exception to this was that BT undertook to fund the whole of its participation and, in addition, contributed the project's share of the national contribution to the European Space Agency (ESA) to provide for the operation of OTS during its use by the project. The principle of partially funding companies to pursue speculative research and development projects was not new, though the proportion was higher than usual. GEC and Logica each contributed 50 per cent of the total costs of their respective participation. What was an entirely new departure was that the DTI also funded to the extent of 50 per cent the SERC sponsored sector of the collaboration, namely the cost of participation by Cambridge, Loughborough, RAL, and UCL. The remaining 50 per cent was provided by SERC. DTI were to be congratulated on their flexibility in making funds available in this way and so making possible an unusually wide collaboration, involving as it did universities, a government sponsored laboratory (RAL), private industry, and a public service organization (BT) in the process of metamorphosis into a private corporation.

Technical direction and day-to-day management of the project was vested in a committee of the principal investigators from each of the participating institutions. Once the project acquired a co-ordinator, this committee met under his chairmanship. The university members of this committee were heads of departments, each of whom had complete technical, managerial, and executive responsibility for his university's participation. By contrast, other members of the committee had varying degrees of responsibility and authority for the project within their own organizations, the remainder being vested in representatives on the project steering committee. The latter consisted of representatives from those organizations with financial responsibility for the project; the steering committee chairman was independent of all the organizations concerned. The chairman of the principal investigators' committee was ex officio a member of the steering committee. Since the source of funding for the participation of the universities was SERC and DTI, there were no university members on the steering committee. The effect of these arrangements within these two committees was to separate the principal investigators' committee from effective financial and some executive participation. At the same time, the steering committee was distanced from the technical issues of the project. While on the whole the management of the project worked well enough, there were occasions when these anomalies caused some difficulties. The experience gained in this project suggests that this is an area where one should seek a more streamlined and uniform arrangement for future endeavours.

The technical design and implementation work of the project was carried out by a series of working groups under the direction of the Principal Investigators' Committee. Not all of these groups existed for the whole of the project. They were formed as necessary when an area of work was identified. They met as often as necessary to accomplish their purposes. Generally, when a group had finished its assigned area of work, it simply ceased to meet. The exception to this was the group responsible for co-ordinating construction of the network: when construction was complete, this group then undertook the operation of the network.

The system of working groups worked extremely well. One of the most important aspects of their operation was that there was considerable overlap among the membership of the groups. This came about largely because of the size of the working membership of the project. However, it turned out to be exceedingly important in expediting the progress of the project. When a number of the groups were meeting every two or three weeks at the height of activity, there were times when it was very important that one group should know in detail what another group had just, perhaps the previous day, decided to do about some issue. Overlapping membership proved to be an exceedingly effective way of achieving this degree of co-ordination at technical working level.

Two further comments relate to items of infrastructure within the project. The first of these was electronic mail. At the beginning of the project, this was just coming into being among the academic participants (though UCL had been habitually using such facilities, particularly in connection with DARPA work). During the last year of the project, its use was regular among all participants and was generally considered a valuable aid. The other item was the means for preparing documents. Towards the end of the project, several of the organizations were passing machine-readable drafts to each other for text editing and comment. While printing of the final result tended to be really satisfactory only at the originating site, the ability to pass drafts back and forth quickly was valuable.

An issue of collaborative management which surfaced in project UNIVERSE was that of intellectual property rights (IPR). This was never fully resolved during the project, partly perhaps as a result of a lack of any particular occasion to force the issue. What did become clear was that the problem is not an easy one to resolve. Since project UNIVERSE, participation by many organizations in the ESPRIT and Alvey initiatives will perhaps lead to satisfactory arrangements for handling IPR; at the time of UNIVERSE there seemed no obvious precedent to follow for such a project.

1.5 Programme and timescales

The whole of the project had its timescales set by the lifetime of the OTS. The operational lifetime of OTS was very uncertain throughout the project. At the beginning of the project in 1980, no formal decision to continue beyond 1981 had been agreed by the ESA council. Technically, OTS was operating perfectly well and it seemed almost certain to remain in operation for several more years. Early in 1981, it was agreed that it would remain operational at least until the end of 1982. It also became clear that in terms of how much gas remained for control purposes, it was probable that it could remain operational until sometime well into 1984. The final decision that OTS would cease service at the end of 1983 was taken towards the end of that year.

Nominally, the project was for three years. Funding for a number of the participants did not begin until late in 1981, notably for the SERC sponsored sector. This meant that for these participants, the project did not formally end until the end of September 1984. Fortunately, a certain amount of internal resources were committed by some of the participants during the first half of 1981 in order that crucial work could be carried out. If this had not been done, there would have been little chance of carrying out the project before the demise of OTS. The principal pieces of development work carried out at this time were preliminary design of the network architecture and prototyping of parts of the ring–ring and satellite bridges.

By the end of 1981 it was known that the critical parts of the network could be made to function. The job of building the network was the task undertaken during 1982. During this time the rest of the programme for the project was developed. The means by which the network was exercised and investigated was organized as a series of experiments, which were undertaken by various subgroups of the participants. These experiments substantially occupied 1983. A full account of this part of the project is given in Chapters 6–9.

During 1982 an additional ingredient was added to the project: namely, that a mobile site should be constructed and public demonstrations of the project should be mounted at three exhibitions during 1983. Since the original project had not envisaged such an undertaking, additional funds of the order of £200k were sought and obtained from DTI for this purpose. These funds covered additional equipment and effort from Logica to organize the demonstrations. The project as a whole regarded these demonstrations with mixed feelings. On the one hand they provided tremendous impetus to make things work by given dates and to that extent undoubtedly helped the project to progress. On the other hand they also diverted a great deal of expert effort at the centre of the project away from the allotted tasks of research and development. A description of the demonstrations is given in Chapter 10.

After the demise of OTS at the end of 1983, a substantial amount of the remaining time was expended in the production of a set of 21 reports embodying a complete description of the project and its results. These reports are listed in Appendix B. The single most important outcome of the project was the demonstration that the design and construction of a wide area network with properties akin to a LAN was possible. And furthermore that one possible means for realizing such a network was by using a broadcast satellite channel for the long-haul segment. In the next four chapters the network, its architecture, and its realization are described.

2. Basic technology of the UNIVERSE network

2.1 Introduction

2.1.1 *General properties of LANs*

During the 1970s a number of high-speed packet-switched communications technologies were developed which have now come to be known as local area network (LAN) technologies. In these systems a number of computers have access to a common shared digital communications channel. These computers, which will sometimes be called stations, may in some fashion obtain brief exclusive use of the channel in order to communicate with other computers attached to the channel. Addressed packets are transmitted on the channel, with each station receiving all packets transmitted, but only reading in those packets addressed to itself. A number of different technologies have been developed using various types of communication channel and different strategies for controlling allocation of the exclusive use of the channel. Well-known examples of these technologies are the *ethernet*, which makes use of contention strategy, the *token-passing ring*, and the *empty-slot ring*.

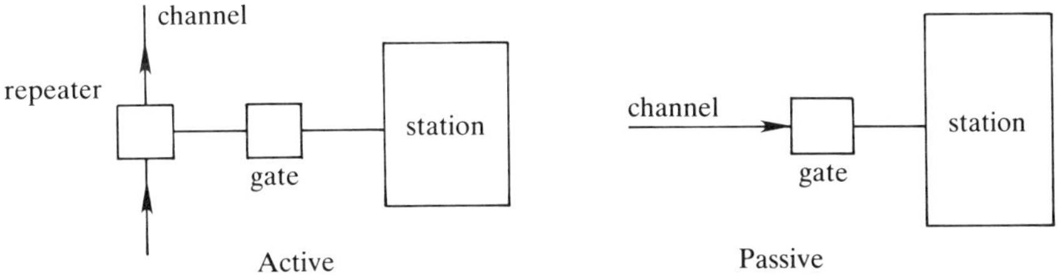

FIG. 2.1 Active and passive modes of operation.

Two categories of channel have been used for LANs: channels in which the receiver is passive and channels in which the receiver must act as a repeater and retransmit the signal to the next station (see Fig. 2.1). In both cases, the switching function in the network is totally distributed; the switching element is essentially a gate or a gate coupled to a repeater. The function of the gate is to look for the start-of-packet sequence, read in the destination address and then read in the whole packet if the address corresponds to that of the host computer.

If useful information is to be transported, stations must transmit in such a fashion that signals from different stations do not collide on the medium. Organization of this access control requires communication between the stations and normally the same channel is used for this purpose. For this signalling between stations to be rapid, the distances between stations must be limited and hence the idea of the LAN, with distances confined to a few kilometres and transit times for signals to cross the network measured in microseconds.

LANs have in common a number of properties that are most desirable for computer networks. These are:

- high-speed transmission. Transmission rates on LANs range from hundreds of kilobits per second to tens of megabits per second.
- short access times. It is possible for a station to access another station on the network within a few average packet transmission times. The Cambridge ring technology, to be described below, gives particularly rapid access times.
- low transit times. The bit transit transmission delay between stations is very short compared to the average packet transmission time.
- high reliability. Reliability, in the sense of bit-error rates and of mean time between failures, is high. Bit-error rates considerably better than 1 bit error in 10^{10} are normal.
- sequentiality guaranteed. Data is always delivered in the order that it is sent. Data may be lost, but there is no possibility of data being delivered out of order.
- congestion control. Congestion is controlled by the network in the sense that there is no point at which the load presented to the network will cause the total throughput to drop. (*Note:* Theoretically, any contention scheme employing a back-off algorithm, for example ethernet, will become unstable at some point if the possible load applied is increased indefinitely by adding stations to the network. In practice, networks operate well below this point.) In addition, these networks do not discard packets in circumstances of heavy load on the network.

LANs also naturally provide broadcast or multicast transmission, that is point-to-multipoint transmission, at zero cost. However, LANs do not provide error correction, nor do they generally provide control of the flow of packets to the receiver. The short delays in LANs come about because of the short distances, the high bit rates, and the absence of store-and-forward delays (apart from the delays of a few bit times in repeaters). These networks may thus be described as 'one-hop' networks, in contrast to conventional 'multi-hop' store-and-forward packet switching networks, where packets 'hop' from switch to switch as well as between switches and hosts.

2.1.2 *Satellite broadcast channel*

With the important exception of the short delay, satellite transmission channels can be used to produce networks with similar properties to those of LANs. In a satellite network, ground terminals transmit microwave signals to a communications satellite positioned in the geostationary orbit above the equator. The satellite acts essentially as a reflector in the sky, receiving signals from a ground terminal, shifting these downwards in frequency and retransmitting them to the ground. The ground area covered by signals from a satellite antenna, that is the area within which signals can be received at reasonable strength, is called the *footprint* of the antenna (see Fig. 2.4). Footprints typically have diameters ranging from hundreds to a few thousand kilometres. By suitable arrangement of the footprints of the receive and transmit antennae on the satellite, which in fact may be the same, a single shared communications channel can be used to provide all-to-all communication between ground stations situated at points spread over a large area of the Earth's surface. Satellites can provide channels having speeds of some number of megabits per second and, using forward error correction, low bit-error rates can be obtained. Clearly sequentiality is guaranteed and since, in a manner similar to a LAN, access must be controlled so that the signal from only one ground terminal arrives at the satellite at any given time, congestion control is also provided automatically. Satellite networks are another example of 'one-hop' networks since there is no store-and-forward action at the satellite.

Satellites have one major disadvantage, which is the long delay time required for signals to travel from one ground terminal to the other terminals in the network. The time taken for a signal to travel from the ground up to the satellite and back to the ground is of the order of one quarter of a second. This is an order of magnitude larger than the delay in a continental-sized ground link. This long delay makes it necessary to adopt different strategies for the control of access to the channel and in order to make efficient use of the channel some type of reservation scheme must be used. The quarter-second delay is noticeable in a number of user applications but is not usually a serious problem. Care, however, must be taken to minimize the number of cross-network interactions involved in any operation.

The LAN technology employed in project UNIVERSE was a form of empty-slot ring technology known as the Cambridge ring. Cambridge rings were installed at all seven of the fixed sites making up the UNIVERSE network and also at the demonstration sites (see Chapter 10). A description of the main features of the Cambridge ring is given in the next section. For communication over the wide area, that is communication between sites, a satellite network based on the use of a shared broadcast channel was employed. A brief description of satellite network technology is given in Section 2.3.

At a number of sites several Cambridge rings were deployed to provide local intra-site communication. These rings and the satellite network were interconnected by computers called bridges to form the UNIVERSE network. A brief description of the function of these bridges is given in Section 2.4.

2.2 The Cambridge ring

2.2.1 *Basic operation*

In an empty-slot ring a number of packet frames of fixed size circulate continuously round the ring. A station wishing to transmit waits until an empty slot appears and then places addressed data into the slot. Each station monitors all passing slots and reads in data from any filled slot addressed to itself. In the Cambridge ring (Wilkes and Wheeler, 1979; Hopper *et al.*, 1986), slots are emptied by the sender and sharing of the ring bandwidth is ensured by ruling that a station cannot continuously use the same slot. So, having emptied a slot, a station must not immediately reuse that slot. Slots are typically very much smaller than the packet sizes normally used for computer networks since the bits comprising the slot must be stored within the ring, that is within the cables and repeaters that constitute the ring. (Special repeaters containing shift registers may be employed to adjust the bit capacities of rings to convenient sizes.)

The Cambridge ring is a bit-serial empty-slot ring running at a bit rate of 10 Mbit/s. Bits are normally transmitted between repeaters over dual twisted-pair lines as 5 V pulses, ordinary four-wire telephone cable often being used to provide the lines. Each pair is fully balanced and in each bit time transitions on both lines indicate a one, while a transition on only one line indicates a zero. Since the ring is constructed out of point-to-point segments, it is not necessary to use the same medium between all repeaters and optical fibre is normally used where a long link is required.

Repeaters are needed wherever a device is to be attached to the ring or where the signal has to be regenerated, for example, at the ends of an optical fibre section. Repeaters pass the bit stream to both their ring output side for transmission to the next repeater and externally for inspection by

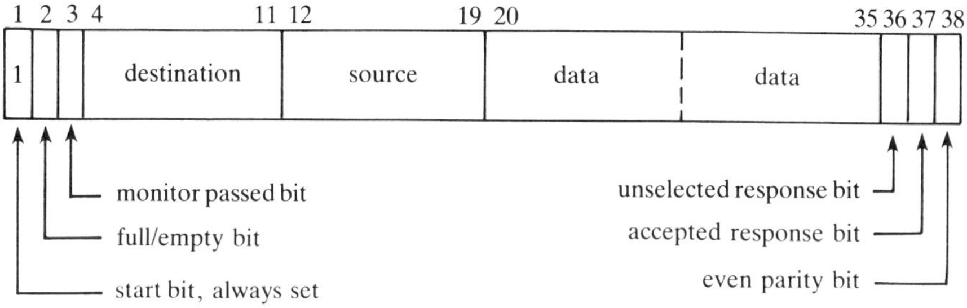

FIG. 2.2 Cambridge ring minipacket format.

an attached device. Repeaters may also substitute external data in place of the incoming bit stream. Repeaters take their power from the ring and are thus not dependent on any attached device being powered on.

Data is transmitted round the ring in *minipackets*. These are 38 bits long and are made up of two data bytes, a destination address byte, a source address byte, four control bits and two response bits. The format is shown in Fig. 2.2. In later, standardized versions of the ring (Sharpe and Cash, 1982), two extra bits, used to aid the implementation of higher level protocols, were added to make 40-bit minipackets. Each ring slot is the correct size to contain one minipacket, the number of slots in the ring being determined by the length of wire and the number of repeaters. Slots are placed nose to tail in a train which circulates round the ring at 10 Mbit/s. There must be a gap of one or more zeros between the head and tail of the train. As has been mentioned previously, the number of slots in a ring is likely to be small; a ring with three or four slots would be quite large.

Wherever a device is to be attached to the ring, a ring node must be inserted. A ring node consists of a repeater together with some additional logic that Cambridge called the station logic. The latter contains an eight-bit address set on a coding plug and whenever the node transmits a minipacket this is automatically added as the source address. It also determines which minipackets are read into the device. Each node also has a select register which may be used to permit reception from any source, from a specified source or from no source. To transmit a minipacket, a host loads the node with the data bytes and the destination address and then gives a transmit signal. The node waits for an empty slot and, when it arrives, it is marked full and the destination, source, and data fields are filled in. The minipacket then passes round the ring and back to the sender.

On its way round it should have passed the destination, in which case the response bits may be marked with one of four possible acknowledgements:

Accepted	[01]	– data and source bytes have been copied into the receiving station.
Busy	[00]	– the receiving station was switched on but was still dealing with a previous minipacket.
Unselected	[10]	– the select register in the receiver had been set to exclude reception from this source.
Ignored	[11]	– response bits unchanged, indicating that the receiving station was either switched off or non-existent.

When the minipacket returns it is always marked empty and passed on. This is an 'anti-hogging' feature that assists in sharing the network bandwidth equally between stations. The action to be taken after a minipacket has been sent and returned unaccepted is defined by the sending host, not by the node. The node does have a back-off feature to prevent excessive traffic being generated by a host repeatedly re-sending a minipacket that is not accepted. The more attempts that are made to send the same minipacket, the longer the node takes to report the failure. The ring nodes used in the project waited one complete ring cycle after the first rejection before transmitting again and thereafter waited fifteen ring cycles between retransmissions.

Each ring has a special station called a *monitor station.* This monitor station is responsible for setting up the framing structure of the ring, including adjusting the gap size to fill the ring and then for maintaining this structure. It deals with the possibility of a slot erroneously being marked full and then left to circulate, thus reducing the bandwidth of the ring. The monitor station carries this out by zeroing the monitor-passed bit whenever a full minipacket passes. The sender must always set this bit so that if the monitor receives a full minipacket with a zero monitor-passed bit, it can note the fault and empty the slot to allow its reuse. The monitor station also makes a number of tests to check continuously on the correct functioning of the ring.

Between the node and the host computer there must be an interface, which is dependent on the type of host involved. A number of different interfaces have been built for the Cambridge ring, ranging from simple ones that generate an interrupt whenever a minipacket has been sent or received, to elaborate interfaces that move data to and from the host's memory by direct memory access (DMA) transfer. The latter usually incorporate an implementation of the ring packet protocol (described below). Using a minipacket interface, it is necessary to make a new transmission request to the node within 3.5 μs of the return of the previous minipacket in order to obtain maximum performance from the ring. Few machines can process interrupts at this speed. However, providing the machine does not often send or receive large blocks of data, the low cost of providing such interfaces can be worthwhile.

A particular interface that was used in many situations in the project was a board, manufactured by Logica VTS and based on an Intel 8086, which gave DMA access to and from Multibus memory. This board also implemented the packet protocol used on the ring and was capable of DMA scatter read and write.

2.2.2 *Basic block protocol*

All communication on Cambridge rings in project UNIVERSE made use of the packet protocol (called basic block protocol) defined by Cambridge for the ring. The packet format consisted of the following elements:

- a header pattern;
- a packet length;
- a port number;
- up to 2048 bytes of data;
- a 16-bit checksum.

Packets are sent round the ring as a stream of minipackets. Operation of the packet protocol proceeds as follows, considering the receiving end first. The receiving station starts with its select register set for reception from any source. Only minipackets containing the header pattern are accepted and on receipt of such a minipacket, the select register is set to accept minipackets only from the source of the header; further minipackets are then awaited. In general, if the packet is expected, a pre-assigned buffer will be ready and the following data minipackets will be read into this buffer. At the end the checksum will be checked and the select register reset to accept from all sources. If the packet is satisfactory, it will be passed on to higher levels of software.

In principle the select register is not required and it would be possible to implement an interface that would multiplex packet reception from a number of sources. No such implementation was used in the project. If the source and port number received at the start of a packet indicate to the receiver that the packet is unwanted for some reason, the receiver may set the select register to reject all minipackets for a short period. The sender will see that the receiver has gone unselected and may take this as an indication that the packet is unwelcome and that its transmission should be abandoned.

A typical implementation of sending a packet will proceed as follows. The header is sent first. If the response is 'busy', it will be tried again; if the response is 'unselected', the sender may try again or attempt to send packets to other destinations before trying this one again. When the header is accepted, the rest of the packet is sent. If during this process a 'busy' response is received, the sender will try again; if 'unselected' is received, it will give up.

Both sending and receiving packets require the use of time constants for timing out certain operations; that is, if there is no satisfactory response to some action before a given time limit, the action is abandoned. There was a reception time-out that applied to the whole packet and this was usually set at one second. On transmission, implementations normally counted events. Typically, after 500 attempts to send a header or 50 to send a subsequent minipacket, the transmitter would give up.

2.2.3 *Data rates*

Although the Cambridge ring operates at a bit rate of 10 Mbit/s, the total data throughput is limited to 4 Mbit/s, on account of the large amount of administrative information carried in each minipacket. The figure 4 Mbit/s is made on the assumption that the gap size is relatively small. Stations cannot use the total bandwidth of the ring, as having transmitted a minipacket a station must wait for its return and then pass it on empty before trying to transmit the next minipacket. In the rings used in the project stations were not able to use the next minipacket either. Thus for a ring with n slots, a single station can only use $1/(n+2)$ of the total capacity, giving a data transmission rate of $4/(n+2)$ Mbit/s. Thus for a three-slot ring, the maximum point-to-point data transmission rate is 800 kbit/s. In practice, few operating systems can sustain even this rate of transmission or reception for longer than a very limited period. In circumstances where a number of stations m $(m > 2)$ are attempting to use the ring as fast as possible, each station will receive a fraction $1/(n+m)$ of the total capacity.

The Cambridge ring has a very low latency as slots pass each station every 4 μs. Rings normally operate at rather low utilization and in these circumstances an empty minipacket will generally be available within a few microseconds. Even in the worst case the wait will only be $4(n+1)$ μs, where n is the number of stations attached to the ring. This low latency makes it possible to consider carrying real-time traffic such as telephony on the ring.

2.3 Satellite networks

In this section, a very brief description is given of the aspects of satellite communication of particular importance to the project. Such important matters as link budgets, modulation and coding schemes, and earth-station construction details are considered to be outside the scope of this book

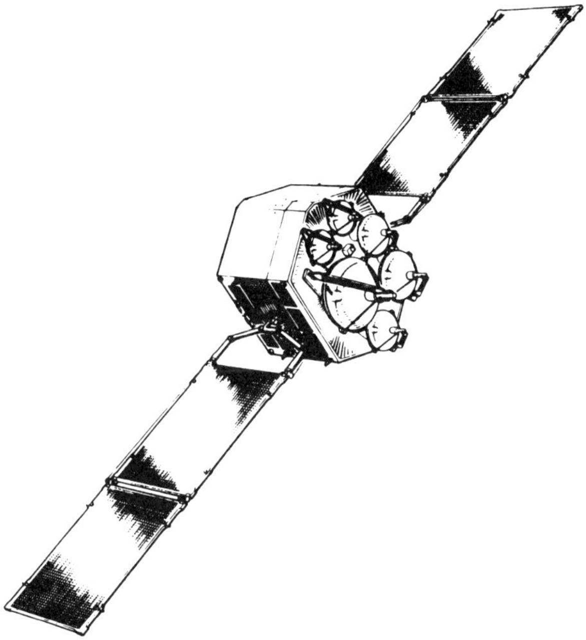

FIG. 2.3 Orbital Test Satellite (OTS).

and readers interested in studying this technology are referred to Pierce and Posner (1980) and Feher (1983). For readers already familiar with the technology, the main parameters associated with the earth stations used in the project are listed.

Communication satellites are currently used primarily either to provide point-to-point digital links, mainly carrying telephone traffic, or for relaying analogue television services. Since satellite transmissions are inherently broadcast, satellites have the potential of providing communications channels suitable for use in the construction of computer networks. In this section the general features of satellite communications systems will be outlined and a description of the particular system used in project UNIVERSE will be given.

Construction of these satellites centres on a cylindrically-shaped box which serves both to house the receiver and transmitter electronics together with the services required by the satellite and to act as a platform for the antennae. One end of the cylinder is used as a platform for mounting the antennae carried by the satellite, the other mounts onto the launching rocket. The cylinder also supports two paddle arms of solar cells, which rotate to face the Sun and provide power for the satellite. Figure 2.3 shows a sketch diagram of the OTS which was used in the UNIVERSE network.

Communications satellites are normally placed in the geostationary orbit. This orbit is unique and it is situated about 36 000 km above the equator. In this orbit the speed of rotation of the satellite matches the rotational speed of the Earth and thus the satellite appears to be stationary, located above some fixed point on the equator. In fact, because of perturbations caused by the gravitational fields of the Sun and Moon, the satellite oscillates about this fixed point. The oscillations are in the shape of a figure of eight with dimensions of the order of 100 km. Geostationary satellites are also subject to drift caused by various forms of friction, and gas jets carried by the satellite must be used to keep it on station. These jets must also be used periodically to adjust the orientation of the satellite so that the footprints from the antennae remain in their required positions. It is generally the supply of this gas which places a limit on the satellite's lifetime.

The communications function of the satellite is to receive signals from a transmitting ground terminal, to amplify these signals and then to broadcast the signals back to the ground after shifting them in frequency so as not to interfere with the incoming signals. On OTS the same antenna was used for both reception and transmission and the signal pattern for this antenna (EUROBEAM 'A') overlaid on a map of Europe is shown in Fig. 2.4. The great attraction of satellite networks will be seen immediately to be the very wide geographical coverage that is available as soon as the satellite is in orbit. Realization of this coverage is, however, dependent on being able to place earth terminals at or close to customer's premises. The seven permanent sites that made up the UNIVERSE network were all located close together in the south of England. However, it would have been quite possible to have carried out the experiment with the sites scattered over most of Europe. For demonstration purposes, a temporary site was constructed in Geneva, Switzerland for the World Telecom '83 exhibition (see Chapter 10).

The electronic subsystem which processes the signal received at the satellite and produces an amplified signal for retransmission is called a *transponder*. Normally a satellite will carry a number of transponders, processing signals in different frequency bands and with different polarizations within a main frequency band. The main frequency bands used by OTS were 14 GHz for reception and 11 GHz for transmission. The transponders on the satellite were wideband, capable of handling signals within a 120 MHz band of the frequency spectrum. In the project the signal rate of the satellite channel was 2 Mbit/s, which only required the use of 5 MHz of this bandwidth. Thus the UNIVERSE network was able to share the use of a single transponder; other experiments used the remaining transponder bandwidth. This method of sharing the use of a transponder is called frequency division multiple access (FDMA).

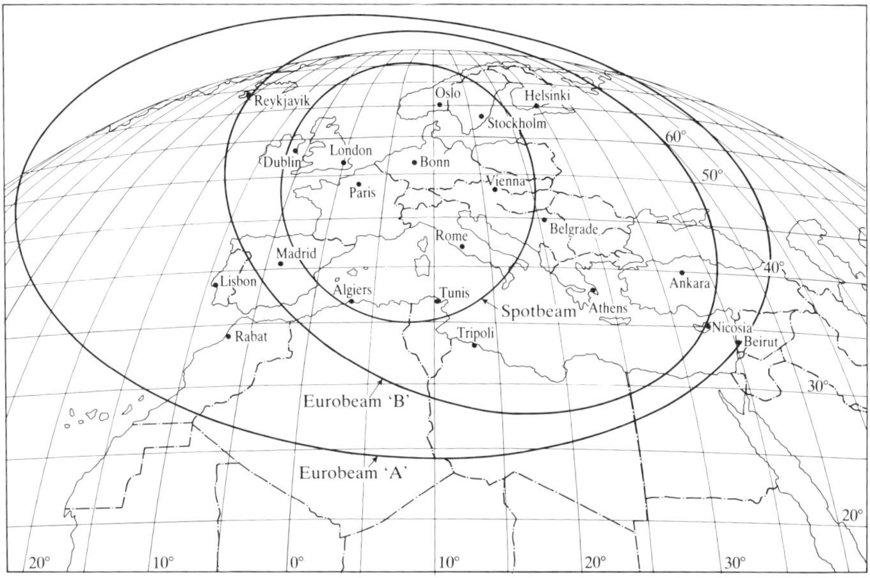

FIG. 2.4 OTS footprint.

For a network of the type envisaged for project UNIVERSE the main requirement for the earth stations was that they should be simple, small, and low cost, thus making them easy to install on the experimenters' sites. It was also necessary that unattended operation should be possible. Antennae having dishes with diameters in the range of 3–5 m can be used satisfactorily at frequencies in the 11–14 GHz range. In the project the ground terminals were equipped with fixed 3 m antennae. Dishes of this size have a sufficiently broad beam to make automatic tracking of the satellite unnecessary and are also quite suitable for roof mounting. The main parameters of the ground stations were a maximum effective isotropic radiated power (EIRP) of 70.9 dBW from an 140 W travelling wave tube high-power amplifier and a receiver G/T (ratio of gain to noise temperature) of 22.5 dB/K. The terminals were driven by 2 Mbit/s burst modems which used binary phase shift keying (BPSK) encoding. The electronics for the ground terminal occupied one six-foot rack.

Data sent over the satellite channel was encoded for forward error correction, using half-rate convolutional encoding and Viterbi decoding (Viterbi and Omura, 1979). This coding produced a very low error rate on the channel of the order of one bit error in 10^9, or better, at the expense of reducing the data rate to 1 Mbit/s. In the project this coding rate was fixed for all types of data; in an operational system it would be necessary to use a number of coding rates to suit different types of data. For example, voice data needs little or no encoding, as the occasional bit error does not spoil the received sound.

The cost of the ground terminals is one of the main factors determining the viability of satellite networks. The cost of the terminals used in the project was of the order of £100k; suitable terminals for this type of network could be purchased in 1986 at a cost of £30k. The annual rental of a 2 Mbit/s bandwidth satellite channel is on the order of $60k in the USA. It is not possible to rent suitable satellite channels in Europe at the present time.

The key problem for a satellite network is how to organize the transmissions from the ground terminals so that they do not interfere with each other. Generally, there are two methods of sharing the basic communications capability of a transponder: by division in the frequency or the time domain. A method of frequency division that is often employed is FDMA, in which a number of frequency bands are defined within the main bandwidth of the transponder. If these bands are suitably chosen, signals in different bands may arrive at the satellite at the same time without interfering with each other. Thus a station may be allocated its own frequency band and transmit continuously in this band without causing interference to signals from other stations. In time division multiple access (TDMA), stations transmit at the same frequency but arrange their transmissions in bursts. The timing of these bursts from different stations is organized to avoid collisions at the satellite. Fixed division schemes are analogous to permanent circuit systems, whereas time-varying division schemes are more akin to packet-based communication. Normally, in operational communications systems, fixed schemes have been used for both FDMA and TDMA. Extensive investigation of variable schemes has been made in a number of research projects; some introductory descriptions of these may be found in Kuo (1981) and Tanenbaum (1981). As already mentioned above, the project used a 5 MHz channel obtained by FDMA in the overall transponder bandwidth. Within this channel, the project used its own form of variable TDMA, with control of the time division being carried out using the satellite channel itself. This imposes a rather long time-constant on allocation decisions, as the time for signals to travel from one ground terminal to the others via the satellite is of the order of 260 ms. Ground channels with low bit rates could have been used for this allocation activity. However, satellite networks are obviously most cost effective in situations where ground links, even slow links, are expensive or difficult to obtain. Hence the strong motivation for using the satellite channel for this purpose.

The allocation method used in the project was a reservation scheme called demand access TDMA. In this scheme one of the terminals is designated as a master terminal and defines a synchronizing time frame by making regular repeated transmissions of a reference signal. This terminal also makes allocations of variable-size time windows within this frame to particular terminals. Terminals periodically transmit requests for transmission capacity based on their current requirements and the master station shares out the available time on the basis of these requests. Within their window allocations, terminals may transmit addressed packets to a number of different destinations. The master station also functions simultaneously as a normal station, and in this role operates in exactly the same manner as any other station. This channel access scheme is described in detail in Chapter 4.

The principal reason for using a satellite network in project UNIVERSE was that this was the only reasonable way, in the period 1980–1983, of obtaining megabit per second links between the sites of the participants. Satellite networks, however, do have a number of advantages over conventional packet-switched networks using ground links. First the satellite gives a 'one-hop' network, with all-to-all communication, which may be continental in size. Thus no equipment is required outside customer sites, apart from the satellite, and customer sites may, in principle, be located anywhere within the large coverage area: on islands, on mountain tops, or in the middle of deserts. The sites may even be mobile, satellite networks being particularly attractive for use with ships. These advantages continue even when high-speed, circuit-switched networks become widely available. The principal disadvantages of satellite networks are the long propagation delay and the cost of ground terminals. Project UNIVERSE served to show that the former is quite manageable and the normal course of electronic technology is already producing solutions for the latter.

2.4 Linking the networks: bridges

Single LANs are limited in capacity in three ways: in the number of hosts that may be attached, in the total data throughput, and in distance and area covered. Thus for a large site, such as a multi-floored office building, university campus, or factory complex, several or even a substantial number of LANs may be required to provide the site network. Two LANs may be interconnected by a computer that is a host on both LANs. These linking computers are often called *bridges* and that name was used in the project. A large site might typically have a number of rings, each with a hundred or so hosts attached, with the rings interconnected by bridges to form the site network. A possible configuration would be to have a number of rings linked together by a *backbone* ring (Bux and Grillo, 1985). Several of the UNIVERSE sites were equipped with more than one Cambridge ring.

The wide area backbone network in project UNIVERSE was provided by broadcast satellite. Each site was equipped with a ground terminal which was connected to one of the site's rings by a satellite bridge. Unlike the local networks, the only hosts attached to the satellite backbone network were satellite bridges. In the case of Logica and University College London, the sites were sufficiently close to allow a cable link to be laid between them. The interconnection was made by having ring half-bridges at each site linked by this cable. The basic transmission rate of the link was 1 Mbit/s and used Logica proprietary signalling mechanism and modems. Short links of this type would clearly be used in operational networks to avoid the cost of additional ground terminals. Thus the UNIVERSE network consisted of LANs and a satellite network; the interconnections were made by ring–ring bridges, satellite bridges, and, in one case, two half-bridges linked by a cable.

The basic function of a bridge is to receive packets from one network, to perform some address mapping, and then to transmit these packets onto the other network. In principle, a bridge could interconnect several networks; only bridges that connected two networks were used in the project. The basic objective of the UNIVERSE design was that the bridges should extend the features of a single LAN or 'one-hop' network into the full interconnected network configuration. Thus, since LANs preserve sequentiality, that is packets are delivered in the order that they are sent, bridges should be designed to maintain sequentiality. LANs do not guarantee reliable delivery of packets nor do they offer flow control services between sender and receiver. Therefore, bridges are not required to offer these services. This turns out to be rather fortunate as the long satellite transmission delay makes the task of providing these services in the satellite bridge very difficult, in fact, hardly sensible. Congestion control is provided by LANs and must also be provided by bridges; and naturally bridges should be designed to minimize transmission delays. In general LANs allow hosts to build all types of service, datagram or call-based, error-free or noisy; they have the potential to support real-time and synchronous traffic, such as voice data. Ideally, bridges should continue to allow these possibilities between remote hosts, that is hosts connected to different LANs, possibly on different sites. The details of how these objectives were tackled and the extent to which they were achieved is the subject of the next chapter, which describes the architecture of the UNIVERSE network.

3. The UNIVERSE network architecture

3.1 Introduction

During the 1970s a number of packet-switched computer network architectures were developed, well-known examples being the ARPAnet developed in the early 1970s in the USA by the Defense Advanced Research Projects Agency (DARPA); DECnet developed by the Digital Equipment Corporation (DEC); Systems Network Architecture (SNA) developed by the IBM Corporation; and the X.25 recommendations developed by CCITT. These architectures took a particular view of communications and the requirements of computer systems appropriate to the technologies of the time. Communication between computers was based on the use of public telephone networks as the basic carriers. These telephone networks had been designed to carry voice in analogue form and when used digitally could only provide communications channels which were slow and error prone. The mainframe computing systems of the 1970s required error-free communications for long periods to provide the services of file transfer and terminal access. A typical computing system would open a few calls per minute but these calls would have a duration of many minutes. Reliability in the transfer of data was the design consideration of paramount importance and took precedence in the design of the network architectures over questions of minimizing delay through the network or of providing simple protocols suitable for use with high bandwidth, less error prone communications channels. During the late 1970s a quite different view of computer systems and communications was built up in a number of research laboratories undertaking the development of distributed computing systems. In these systems a number of smaller computers are joined together by a communications medium to form a collaborative system that can carry out the full range of computing functions. LAN technologies, with their high bit rates, low error rates, and small delay, have normally been used for the interconnection of the component computers of these distributed computing systems. Computers in such systems may open many connections per second to a number of different machines which are collaborating to provide a particular service. LANs are able to support this transaction style of communication but can clearly also carry the traditional computer-to-computer traffic of file transfer and terminal access associated with slow networks. A fully integrated services network would also have to carry voice and video alongside the computer generated traffic. Again local network technologies have the potential, as yet unproven, for carrying this type of real-time traffic. Thus LAN technologies are candidates for providing fully integrated services in the local area. The principle aim in the design of the UNIVERSE network architecture was to extend the features of a single LAN into the wide area, and hence to investigate the possibilities of constructing a wide area packet network capable of providing fully integrated communications services.

3.1.1 *The need to interconnect LANs*

LANs have the highly desirable characteristics of high transmission rates, very low bit-error rates, low delays, and of giving very direct communication between end-users. However, they have severe limitations both in the geographical area and in the number of devices that can be serviced by a single LAN. Generally, the practical limit to the number of computers that may be connected by a single ring is of the order of one to two hundred. Thus, even within a single site there is a natural requirement to connect individual LANs to each other in order to accommodate the numbers of computing devices that may require interconnection within that site. As has been described in Section 2.4, this interconnection between LANs is made by computers called *bridges* which are stations attached to both the LANs that are to be linked. The principal consideration in the design of bridges is that they should make the interconnected set of LANs behave, as far as possible, in a manner similar to that of a single LAN.

Of equal importance to the need to interconnect LANs for the purpose of covering a single site is the requirement to interconnect sites. There are two technologies that meet the requirement for long-distance transmission at high bit rates: broadcast satellite communication and high-speed, point-to-point ground links. Both these technologies were used in project UNIVERSE. High-speed switched circuits were not available to the project and are unlikely to be available until the 1990s. Although high-speed point-to-point circuits are now generally available, at the time of the project they were not, and the project could only make use of the single, specially installed, short cable link between Logica and UCL. The main wide area network technology used to interconnect the sites of the UNIVERSE network was TDMA use of a broadcast satellite channel. This satellite network was joined to a LAN network at each site via a satellite bridge computer. Again, in the extension from a single site to a multi-site set of interconnected LANs, the main design consideration for the network architecture was that the complete network should appear to the users to be a single LAN.

3.1.2 *Styles of network service*

Communication between hosts on a LAN may be connectionless, by datagrams, transaction oriented, or over virtual circuits. Datagrams are units of data which enable hosts to communicate with each other without the network imposing any structure on the communication. The network transports each datagram independently. Transaction protocols are frequently used on LANs to support distributed computing applications. In a transaction, a client sends a request to a server in a single data unit of communication. The server responds by sending a reply data unit back to the client, at which stage the transaction is complete. Virtual circuits on a packet network are connections between pairs of hosts that allow the exchange of streams of data in each direction between these hosts. Once established a connection may last for an indefinite period. Each of these modes of communication has many variations in the degree of reliability, flexibility, and quality of service which may be presented. In some architectures, the network may provide only a primitive form of the facility and it is a host matter to construct the more advanced levels. In other network designs, only a very full form of each style of service may be accessible. Furthermore, in most current networks, only a subset of these forms of network access is available.

3.1.3 *Network layers*

It has been found convenient to describe the organization of communication in computer networks in terms of a number of layers or levels of activity, with each level offering services to the layer above. At the lowest level, questions concerning the transmission of bits on a physical medium are addressed, whilst the highest levels deal with matters associated with particular network applications, such as file transfer or terminal access. A very well-known structure for modelling computer network architectures, called the Reference Model for Open Systems Interconnection (OSI), has been developed by the International Standards Organization (ISO, 1983; Day and Zimmermann, 1983; Linington, 1983). This OSI model has seven layers, of which the lower three are concerned with the mechanism of conveying data across the network and describe the interactions between hosts and the network components, whilst the upper four layers deal with end-to-end interactions between hosts in a network-independent way. In terms of this model, the methods of transmitting bits on the Cambridge ring and over the satellite network, briefly described in Chapter 2, constitute the lowest level, the physical layer, of the UNIVERSE network architecture. At the next level, the data link layer, the task described by the model is the point-to-point transmission of packets of data. For the Cambridge ring, this layer divides into two parts, that dealing with transmission of minipackets and a sublayer above this, the basic block protocol, which amalgamates minipackets into packets (see Section 2.2). The method used to transmit packets across the satellite segment is described in Chapter 4. The third level, the network layer, is the main subject of this chapter.

3.1.4 *Informal description of architecture*

The UNIVERSE network architecture supported three types of network access: datagram, transaction, and a lightweight form of virtual circuit. Above these protocols a number of different end-to-end protocols may be used for various applications. Datagrams carry the full UNIVERSE address of the source and destination hosts and are routed hop-by-hop as they pass through bridges. The network does not guarantee that datagrams sent from the same source to the same destination will arrive in the same order as they were sent. Datagrams were used principally in the network for measurement and monitoring purposes. The form of transaction protocol provided in UNIVERSE is of the primitive variety. A client sends a request packet to a service and receives a reply packet or packets in return. Error recovery is by time-out: in the absence of a reply, the client simply repeats the request. The request packet is treated by the bridges in a similar fashion to a datagram but with the addition that, as the packet proceeds to its destination, the bridges create a return path for the reply packet or packets. Sequentiality is maintained in this reply path but no error recovery is provided for corrupted or lost packets. The reply path is closed in the bridges by time-out.

The form of virtual circuit supported by the architecture is a path that carries data sequentially across the network but which provides no mechanisms for recovery from error or for control of data flow. It has been given the name *lightweight virtual circuit* (LVC) in distinction to the *heavyweight virtual circuit* typically supported by packet networks, which have built-in facilities for both error recovery and flow control. The UNIVERSE LVC provides two functions, the delivery of data across the network via a number of bridges in a fashion compatible with the same operation on a single LAN and the maintenance of sequentiality. Thus those blocks travelling along an LVC which arrive safely at the destination, arrive in the same order as they were sent. After use, an LVC is removed by time-out rather than explicit closing. To make use of such LVCs, different end-to-end transport protocols appropriate to particular applications may be used. Thus for file transfer and terminal access, error-free transport is required, whereas for voice and other real-time traffic minimum delay is essential and quite different transport protocols must be used. Both well-established transport protocols, such as Cambridge byte stream protocol (BSP) (Needham and Herbert, 1982; Hopper *et al.*, 1986), and protocols developed within the project, for example the transport protocol for packet voice (see Chapter 8), were carried on the network.

In addition to the network bridges described above, nameservers also play an essential part in the UNIVERSE network architecture. Each ring in the network has its own nameserver, providing clients with a service to map names into network addresses. Clients on the ring have only to remember service names rather than addresses, and make use of the nameserver to bind particular service names to host addresses on that LAN. The UNIVERSE architecture extended this binding to remote services, that is services on a different LAN, by providing a route to that service, established at the time of the name lookup.

Using the combination of bridges and nameservers, it was possible to design the protocols for the UNIVERSE network in such a way that hosts were presented with the same view of communication regardless of whether the communication was on a single LAN or across the full network. The details of how this is achieved are the subject of the next two sections of this chapter. In addition to the two-party communications services described above, advantage was taken of the broadcast nature of the satellite transmission to provide point-to-multipoint communication services within the network architecture. These services are described in Section 3.5.

3.2 Protocols on a single ring

The Cambridge ring is a slotted ring in which a small, constant number of slots circulate continuously. Minipackets of data fit into these slots and can be sent between hosts on the ring. These minipackets contain source and destination addresses of hosts on the ring and a small amount of user data. In Chapter 2, a protocol to allow blocks of information to be sent between hosts in a sequence of minipackets was described. This basic block protocol gives the ring a datagram carrying service, with the source and destination addresses of the datagram carried in each minipacket and the length and a checksum carried in the first and last minipacket of the sequence. Used with this basic block protocol, the service provided by a Cambridge ring becomes conceptually very similar to that provided by other types of LAN, such as ethernet or token ring.

The basic block protocol also provides a port number field, to allow subaddressing within the receiving host. Typically a host might have a ring handling task responsible for receiving from and transmitting to the ring. This task would receive requests from other tasks to receive on particular port numbers. When data arrives addressed to a port associated with a reception request, the received data is passed on to the requesting task. It is convenient to distinguish three modes of use associated with ports: *public*, *well-known*, and *private*. A *public* port will receive from any client

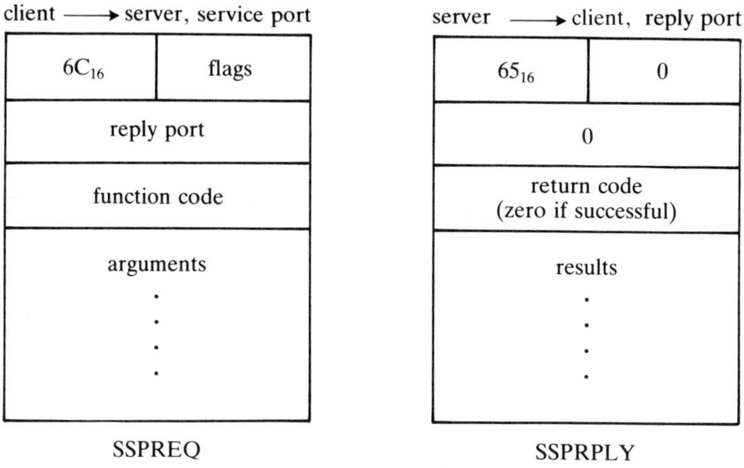

FIG. 3.1 Single shot protocol (SSP) formats.

on the ring. It may or may not be permanently allocated. A *well-known* port is a public port which is permanently allocated and is registered in the nameserver. Well-known ports are generally available for use on a continuous basis. A *private* port, on the other hand, is allocated dynamically, as the need arises, when a connection with a specific host is to be established and its use is confined to the particular connection.

3.2.1 *Single shot protocol*

Above the basic block protocol there are two commonly used protocols: a transaction protocol, called the single shot protocol (SSP), and the byte stream protocol, BSP, already mentioned above. The latter is a network-level stream protocol providing error-free, flow-controlled connections between hosts. In the single shot protocol, an SSP request (SSPREQ) is sent in a single basic block from a client to a service and is answered by the service sending back an SSP reply block (SSPRPLY). The formats of these blocks are shown in Fig. 3.1. The port to which the request is sent is a well-known port on the server. The SSPREQ contains a field giving a reply port number; this is a private port on the client and it is to this port that the server must address its reply (SSPRPLY). The SSP is not a reliable protocol in the sense that, having issued an SSPREQ, a client is not certain to receive a corresponding SSPRPLY. Some time-out must be associated with the activity; if no reply has been received within this time-out period, the client must assume that no reply will be forthcoming and act accordingly, either by retrying the SSPREQ or by following some fault procedure.

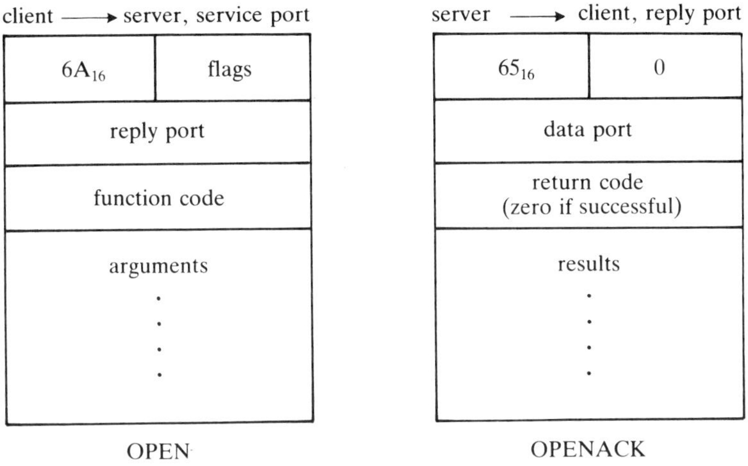

FIG. 3.2 OPEN formats.

3.2.2 *Lightweight virtual circuit*

The network-level stream protocol, BSP, distinguishes two phases: one for establishing a connection between two hosts and a second for the provision of error-free and flow-controlled exchange of data between the hosts using this connection. As far as the architecture of the UNIVERSE network is concerned, only the procedure for connection establishment is relevant. Connection establishment is accomplished by an exchange of blocks. The formats of the blocks exchanged, the OPEN and OPENACK blocks, are shown in Fig. 3.2, and will be seen to be similar to those of the SSPREQ and SSPRPLY. The difference is in the use of the reply port, and the addition of a data port in the reply block, the OPENACK block. The data port is the port private to this connection, which the initiator must use when subsequently sending data over the connection. In the case of an SSP transaction, the reply port is only active until the reply has been received. In the case of an OPEN/OPENACK exchange, the reply and data ports, each host transmitting on one and receiving on the other, define the connection and remain active for the duration of the connection. The receiver of an OPEN block may reject the request to open a connection by sending an OPENACK block with a suitable return code. Again, the OPEN block is sent to a well-known port, while the reply and data ports are private. As with an SSPREQ, some time-out must be associated with an OPEN request. Once the OPEN/OPENACK transaction has been successfully completed between two hosts, a symmetric connection exists between these two hosts. End-users may use this connection, once established, in any way they choose. In particular, the data phase of BSP, referred to above may be used (Needham and Herbert, 1982; Hopper *et al.*, 1986). This provides procedures for exchanging data across the connection in an error-free and flow-controlled fashion. In the UNIVERSE architecture, these procedures are considered to be end-to-end functions and as such are not part of the service offered by the network.

3.2.3 *Nameserver*

One further feature of single ring operation is relevant to the development of the UNIVERSE architecture. On the Cambridge ring, whatever activities a host is going to engage in, it is common practice for it to find the address of the destination host via a name-lookup service. This name-lookup service is provided by a host on the ring called a *nameserver*. Hosts on the ring use names, rather than addresses, for referencing services and the only address normally bound into host software is that of the nameserver. To find the address associated with a particular service name, hosts engage in an SSP transaction with the nameserver, passing it a name and receiving in reply, a station address, port number, and function code. These numbers are the current ring address of the service. The nameserver may also give an indication that the service requested is slow, that is the time taken by the service to process an SSPREQ or an OPEN is long. This indication may assist the client to choose an appropriate time-out.

3.3 The UNIVERSE network services

The UNIVERSE architecture supports three types of network service, a connectionless datagram service, a service for conducting transactions, and a service to provide LVCs between clients on the network. These correspond to the equivalent services on a single ring, with the addition of an explicit datagram service. (Note that, for a single ring, there is no specific datagram service, as the basic block transmission protocol serves to provide this facility.) However, in contrast to a number of network architectures, the protocols for the transaction and LVC services are not built on top of the datagram protocol and the three services must be considered to stand side by side.

3.3.1 *Datagram protocol*

The UNIVERSE datagram is a new type of basic block containing UNIVERSE global addresses for the destination and source hosts. The format of this global address is shown in Fig. 3.3. The site number corresponds identically with a satellite earth station number, so that physically separated sites joined by ground links are considered for addressing purposes as single

8 bits	8 bits	16 bits	16 bits
site	subnet	host	port

FIG. 3.3 UNIVERSE global address.

sites. The subnet number is the address of a ring or other local net within a site and the host number corresponds to a station on a ring. This form of addressing together with the use of a broadcast satellite channel for inter-site communication eliminates the need for consideration of inter-site routing problems; if traffic is to be sent off-site then it must be sent via a single bridge, usually the satellite bridge. Routing within a site was fixed and was built into some of the bridge implementations; others obtained routing information from a static table in the nameserver.

The format of the UNIVERSE datagram is shown in Fig. 3.4. For a client system wishing to send datagrams to a remote service, the sequence of events proceeds as follows.

(1) *Name lookup* The client carries out a transaction with its local name-lookup service, sending an SSPREQ to the nameserver with the name of the required service in the data part.

(2) *Name-lookup reply* The nameserver replies with an SSPRPLY giving the station number B_1 of a bridge on the ring and a port number P on that station, together with the global address of the destination system providing the requested service. The client may cache this information for subsequent use.

(3) *Datagram sent* The client then constructs and transmits datagrams containing the global address supplied, using B_1 and P as the address for the first hop in the path to the destination.

(4) *Bridge forwards* On receipt of the datagram the bridge uses the global address to construct the station number B_2 and port number P of the next bridge in the path, substitutes these in the appropriate fields of the datagram and forwards the datagram towards its destination. The port P on which bridges receive datagrams is a well-known port and is built in, so that bridges do not consult the nameserver to find it.

(5) *Forwarding to the destination* The last bridge in the path will use the station number and port number in the global address to forward the datagram to its final destination.

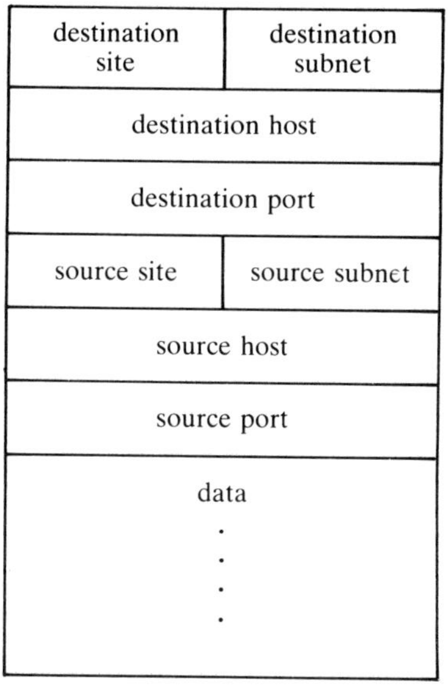

FIG. 3.4 UNIVERSE datagram format.

It would have been possible to have made this type of datagram service the basic carrier mechanism of the UNIVERSE network as has been done in the DARPA Internet architecture (SRI, 1982*a* and 1982*b*). This was not done and in fact datagram communication was used on the network mainly for purposes of network monitoring and was not made compatible with datagram communication on a single ring.

3.3.2 *SSP and LVC path establishment*

The transaction and virtual circuit services in the architecture are based on the use of paths that cross the network via a number of bridges. These paths are logically transparent to the user, in the sense that the user transmits packets to the first bridge in the path as if this bridge is the destination host. Paths are not physically transparent since the delay time for a packet to traverse the whole path may be much larger than the normal time for a hop on the client's local net. Paths are created when a client sends an SSPREQ packet or an OPEN packet. For the transaction service only a return path is set up; for the virtual circuit service both forward and return paths are set up. In both cases paths are closed by time-out, the time-out period normally being shorter for paths set up for transactions.

The procedures used to set up a virtual circuit between host systems on different subnetworks are described in detail below. The basic scheme used for establishing paths is to send an *opening* packet across the network which allows the bridges to set up a path in the reverse direction. In order to be routed across the network, this *opening* packet is transformed to contain global addresses for routing between bridges. Paths are represented in bridges as mappings between a port number on which the bridge will receive packets following the path and a destination host number and port number pair that the bridge will use to forward these packets across the next hop in the path. *Opening* packets have a field containing a reply port number. When a bridge receives an *opening* packet, this reply port number is stored together with the immediate source host of the packet and a new port number on the remote side of the bridge is allocated for the mapping of the return path. The reply port number field is changed to this new reply port number and the *opening* packet is forwarded on its next hop. The bridge now has a mapping defining a path in the reverse direction. This procedure has been likened to shooting an arrow with a string attached. When the arrow has reached its destination, the string can be followed back to the source. It would have been possible to have used SSPREQ, OPEN, and OPENACK blocks with global addresses appended as the *opening* blocks for establishing paths across the UNIVERSE network; however, in this case compatibility with single ring procedures would not have been maintained. By incorporating the nameserver in the path establishment process it was possible to retain full compatibility. In the case of a service being remote, the nameserver responds with a public port, called an *address insertion port*, on a local bridge which a client will use in the same way as a well-known port on a server. An *opening* block sent to an address insertion port will have the global address of its destination inserted in the case that it has to be forwarded to another bridge. This global address is provided to the bridge by the nameserver as part of the name-lookup process.

For a client, located at station C, wishing to set up a connection to a remote service called *RSERV* located on a machine at site S, subnet N, with station number H, and well-known port P_w^h, the sequence of events, shown diagrammatically in Fig. 3.5, is as follows.

(1) *Name lookup* The client sends an SSPREQ to the nameserver on its ring with the name *RSERV* in the data part.

(2) *Address insertion request* The nameserver finds from its tables that *RSERV* is on a remote subnet and makes a request to a bridge, station B say, on the local ring, for an address insertion port to site S, subnet N, station H, and port P_w^h to be set up on the bridge. This is sent to the bridge as an SSPREQ (on a well-known port for use by the nameserver).

(3) *Address insertion reply* The bridge allocates a port number, say P_a^b, and associates the address with this port. The port number is then returned to the nameserver in an SSPRPLY.

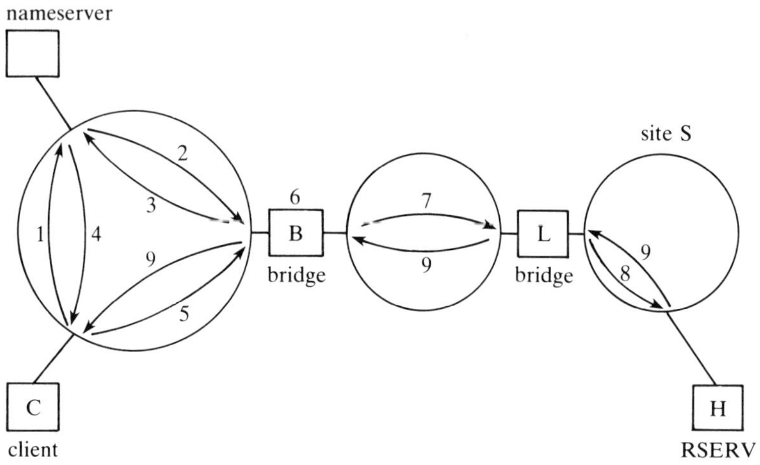

FIG. 3.5 Call establishment through bridges.

(4) *Name-lookup reply* The nameserver replies to the client giving station B, port number P_a^b as the service address.

(5) *OPEN sent* The client sends an OPEN block to the bridge at station B, on port P_a^b, giving a reply port P_r^c.

(6) *Conversion to BRIDGEOPEN* The bridge receives the OPEN block on the address insertion port, and changes the block into a BRIDGEOPEN block which contains global addresses. The format of the block is shown in Fig. 3.6. The destination global address is set to site S, subnet N, station H, and port P_w^h.

(7) *Forwarding BRIDGEOPEN—reverse path set-up* The BRIDGEOPEN block is forwarded to the next bridge in the route towards the destination (using a port well-known among bridges), each bridge making any required routing decision using the global destination address, until it arrives at a bridge attached to the correct subnet, N. The routing mechanism used here is the same as that for datagrams (Section 3.3.1). As the BRIDGEOPEN passes through each bridge, a new reply port is allocated. This maps to a station number, port number pair, where the station number is the station which sent the BRIDGEOPEN (or the original OPEN in the case of the first bridge) and the port number is the reply port contained in the received BRIDGEOPEN. At each hop the reply port is changed to the new reply port before the BRIDGEOPEN is sent on. Thus a reverse path is set up as described above. The mapping associated with the new reply port also has some state variable associated with it to indicate that an OPENACK is expected on the reply path. A timer is set for this reply path.

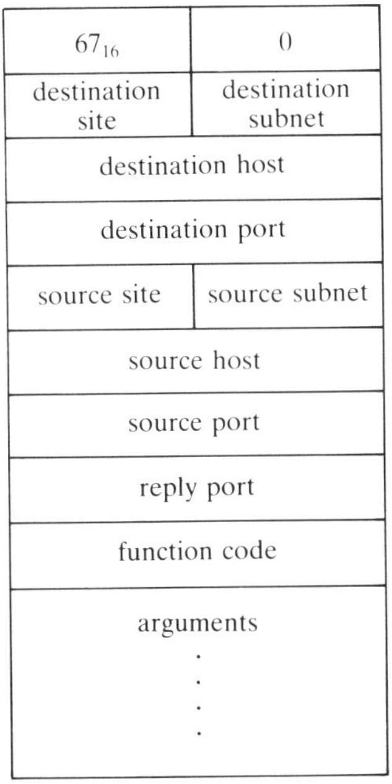

FIG. 3.6 Format of BRIDGEOPEN.

(8) *Conversion back to OPEN—delivery to the destination* When the BRIDGEOPEN arrives at the last bridge, station L say, on the final subnet, N, it is changed back into an OPEN block and sent to station H, port number P_w^h. The reply port, P_r^l say, in this OPEN is one allocated by the last bridge.

(9) *Return of OPENACK—forward path set-up* The service *RSERV* sends an OPENACK back to port P_r^l, station L, quoting a data port, say P_d^h, that it will use for the connection. The bridge allocates a data port for itself and modifies the data port in the OPENACK. It stores this data port to make the path in the forward direction. The bridge then returns the OPENACK along the reply path, no insertion of global addresses being necessary as this reply path for the OPENACK is already set up. (In consequence of this, there is no need for a different format or type of block when returning an OPENACK between bridges.) The bridge also modifies the mapping on the reply path to indicate the passage of the OPENACK and that no further examination of blocks on this path is necessary. The OPENACK then proceeds along the reply path, hop by hop, with bridges creating the forward path as it passes. Finally the OPENACK is delivered

to the client with data port, P_d^b say, allocated by the bridge at station B; at this point the connection has been set up.

(10) *Circuit shut down* Each bridge notes the use of the reply and data ports. If no traffic arrives on these ports for a fixed time, they are (independently) deallocated. The port space at each bridge is large, so that deallocated port numbers can remain dormant for a long time.

A similar procedure to that described above is used for the transaction service, with an SSP request (SSPREQ) taking the place of the OPEN block and a BRIDGESSP block being generated by the first bridge. However, in this case no forward path is constructed, the bridges simply passing on blocks sent along the reply path. In addition, the value used to time out the path is smaller.

Once paths have been established, they may be used in any way that the connected clients choose. The bridges maintain sequentiality on these paths, that is packets proceed along paths in the order in which they are sent, but no guarantee is given that packets will not be lost or corrupted. No explicit facilities are provided to prevent the transmitting end of a path over-running the capabilities of the receiving end.

Using the procedures described above, a client is entitled to cache the result of a name lookup, even though this may be an address insertion port on a bridge, rather than a well-known port on a server. This is allowed because address insertion ports share with well-known ports the property of being public, that is they will receive from any host system, not just from a single station. Thus the strategy in the bridges is to refresh the timer associated with an address insertion port, whenever an OPEN or SSPREQ is received on it, or whenever an address insertion request is made by the nameserver that can be satisfied by an existing address insertion port. In this way, paths leading to popular well-known ports (such as a command port on a fileserver) will virtually never time out. Thus, although clients must be prepared to look up the name of a service if an SSP or OPEN transaction fails, they will generally not have to suffer the overhead of doing this on every connection. This is especially important when a series of transactions is undertaken.

3.3.3 *Time-outs and congestion*

Two further problems associated with network services remain to be covered: these are time-outs and congestion in bridges. As with single ring operations, clients must be able to associate suitable time-out intervals with SSPREQs and OPENs, but now the additional factor of possible network delays must also be taken into account. The solution adopted was for the nameserver to give additional information about the location of services when replying to name-lookup requests, in fact indicating whether the service requested was local or remote.

To deal with congestion in bridges, the technique adopted was a combination of the use of large buffer memories, the blocking of all input, and, finally, of selectively discarding packets. No means of blocking the establishment of new connections through bridges to limit traffic flow was implemented. The use of LVCs implies that there is no method for bridges to limit the flow of packets on individual connections and thus bridges could only protect themselves by selectively discarding packets belonging to the most heavily used connections. In the special case of the Cambridge ring, it is in fact possible to block input to a bridge from a ring because of use of acknowledgements at the minipacket level. In practice, this was done, though not selectively.

3.4 Names, addresses, and the name-lookup service

The relationship between names, addresses and routes in computer networks is the subject of a number of papers in the literature, see for example Shoch (1978), PSS User Forum (1980), and Saltzer (1982). The terminology and structure used in the UNIVERSE architecture follows fairly closely the model proposed by Shoch in which

> A *name* is what one seeks, an *address* is where it is, and a *route* is how to get there.

The term *name* is used to mean the name of a service (or client) or the name of a host (or station); it corresponds to the use of the word *title* in PSS User Forum (1980). The term *address* denotes either a host address, which is the name of a network attachment point, or a service address which is a doublet of host address and port number, the latter providing information for distinguishing between a number of services offered by a single host. The term *route* denotes a path between two processes in different hosts.

3.4.1 *Role of nameserver*

In the UNIVERSE architecture host systems attached to the network only know about names and it is the nameservers, which are systems that are part of the network, that are used to bind names to addresses and routes. Each LAN in the UNIVERSE network has a nameserver and the local address of the name-lookup service on this nameserver is the one address that is bound into the software of network hosts. The nameserver provides all bindings between names and addresses and in addition performs the first step in the binding of an address to a route as described above. Thus the architecture allows situations in which there may be a large number of nameservers in the network; specifically, in UNIVERSE, there was one for each LAN belonging to the network. These nameservers deal with a large number of names, one for each service on the network, and clearly some strategy for name management must be adopted.

3.4.2 *Domains and name management*

This was not an area that was investigated in any depth in the project. The solution adopted was to regard the whole of the UNIVERSE network as a single naming domain and then to partition the corresponding name space into a number of subdomains, allowing interpretation of names at any point on the network, but giving each organization the possibility of managing its own subdomain of the name space. A full UNIVERSE name is thus made up of a subdomain name followed by the name within the subdomain, each name within a subdomain being unique. The syntax of a UNIVERSE name was

$$\langle\, \textit{UNIVERSE name} \,\rangle ::= \langle\, \textit{name of subdomain} \,\rangle * \langle\, \textit{local name} \,\rangle\,,$$

where ⟨ *local name* ⟩ is the name of a machine or service within a subdomain. Table 3.1 lists the names used for each of the subdomains in the network and the associated organization. For example, there was a service called *sitelog* at every site: the full UNIVERSE name for the one at Logica was *logica * sitelog*. Each nameserver, and thus each subnet, has a home subdomain; all the nameservers within a subdomain have all the names used in that subdomain listed in their tables. When a name belonging to a foreign subdomain is presented to a nameserver, the nameserver may have to interact with a nameserver in that foreign domain in order to ascertain the address of the named service or host. This nameserver-to-nameserver interaction across the network is carried out by use of a normal SSP transaction. The nameserver performs on its own behalf the necessary address insertion interaction with the bridge, in the same way as for any other client. Obviously, each nameserver within a subdomain must have built-in addresses for at least one nameserver in each of the other subdomains. Nameservers cache the addresses of foreign names so that frequent references to foreign names do not continually incur the delay of a cross-network interaction; this is particularly important when the delay of a satellite hop is involved.

For purposes of naming, the UNIVERSE network was considered as a single network having a single unified naming domain. The network was, however, connected to other networks, in particular the JANET network (see Section 3.6), via network gateways and the nameservers provided facilities for name lookup of services located outside the UNIVERSE domain. For a request asking for a service located outside the UNIVERSE network,

TABLE 3.1 Subdomain names

Subdomain	*Organization*
btrl	British Telecom Research Laboratories
camb	Cambridge University
gec	GEC Marconi Research Centre
logica	Logica Ltd
lut	Loughborough University of Technology
ral	Rutherford Appleton Laboratory
uclcs	University College London Department of Computer Science
show	Mobile demonstration

the nameserver initiates a connection to the appropriate gateway and returns a public or well-known port in the usual way; in addition, it returns a name which may be sent to the gateway to allow it to continue the process of establishing a route to the service. This facility of the nameserver was a general one but was only used in this particular context. Further details of this are given in Chapter 5.

3.5 Broadcast services

Both the LAN technology and the satellite network technology used in the construction of the UNIVERSE network are inherently broadcast technologies, that is data transmitted by a station attached to a LAN can be received by all stations attached to that LAN, and similarly data transmitted to the satellite from one satellite earth station can be received by all satellite stations. In order to provide facilities for applications to make use of this broadcast capability, a broadcast architecture and supporting protocols were designed and implemented during the project. The scheme used was to define a set of broadcast access points within the network, with broadcast communication being undertaken between these points. Thus a station on the network wishing to transmit in a broadcast fashion sends its data to the nearest broadcast access point. This data is then transmitted to all the broadcast access points in the network and stations wishing to receive the data collect this on a one-to-one basis from their nearest access point (see Fig. 3.7). Broadcast channels are defined between the access points, so that a number of independent broadcast activities may be carried on simultaneously within the network. The access points provide a multicast reception service for each channel, allowing stations to indicate to their local access point that they wish to receive all data transmitted on a given broadcast channel.

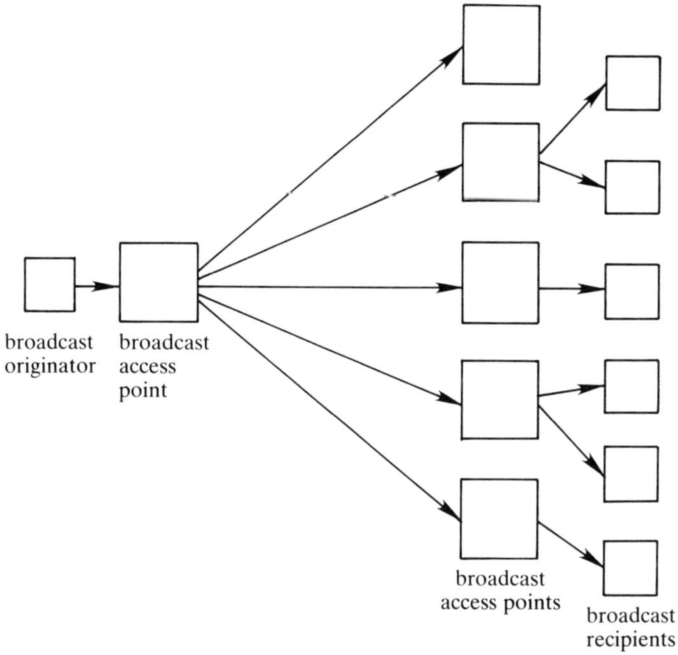

FIG. 3.7 The broadcast architecture.

The main application of broadcast services studied in the project was to file distribution (Chapters 7 and 9). In this application the broadcast technique can be used to minimize the use of satellite transmission capacity when large files are transmitted to a number of sites. Other applications might include simultaneous updating of distributed databases, broadcasting status information (Chapter 7), and multi-party activities such as teleconferencing.

In the UNIVERSE network, only the satellite was used in a physical broadcast mode and each site was provided with a single broadcast access point. In an operational network this arrangement would be reasonably satisfactory as it is inter-site, rather than intra-site, data communication that is expensive. Originally it was planned that the broadcast service access point for a site would reside in a separate machine on the same ring as the satellite bridge, but in the event it was found convenient to incorporate the broadcast service within the satellite bridge itself. The broadcast transmission between the access points was implemented using a broadcast protocol over the satellite network. Details of this protocol are discussed in the next chapter.

3.5.1 *Network broadcast protocol*

A client wishing to transmit data by broadcast starts by opening an LVC to the local broadcast access point. This is done in the normal way starting with a nameserver transaction, which indicates a host number and port number associated with the service, followed by the transmission of an OPEN request. The OPEN request carries a parameter defining the broadcast channel that the client wishes to use. A client may use a *well-known* channel, in which case the channel number is supplied by the nameserver (in the function code) and is quoted by the client in the OPEN request. These well-known channels have the same properties as well-known ports: they are permanently allocated, registered in nameservers, and may be used by any client transmitter. Alternatively, a client may quote a channel number that it wishes to use, in which case the channel must already have been allocated, or the client could ask for a new logical channel to be allocated. In the latter case the channel number would be notified to the client in the OPENACK block that is returned. In all cases, this OPENACK will also return to the client a port number on the access point to which all subsequent blocks for transmission are to be sent. The connection between transmitter and broadcast access point is the same as an LVC in terms of establishment: in particular, no special processing is required in intervening bridges in case the transmitter and access point are separated by local bridges. However, once established, data is only ever sent in the direction from transmitter to access point.

The allocation of channel numbers has to be co-ordinated between the set of broadcast access points. The scheme used for this co-ordination was to define one of the access points as the 'master' point in charge of allocation. Other points then negotiate with this point in matters of channel allocation. Two types of channel are made available, protected and unprotected. On a protected channel only one transmitter is allowed. An unprotected channel allows several originators to transmit on the same channel and these originators may make use of different access points. Well-known channels are clearly of this unprotected type.

The access points also provide a multicast reception service. A client wishing to receive blocks on a specific logical broadcast channel opens a route to its local access point specifying the channel number required in the OPEN block. Thereafter the client will receive any broadcast blocks which arrive at the access point on the appropriate channel. If several clients have opened routes to receive on the same logical channel, the access point will service these clients by repeated point-to-point transmissions. When private channels are used in a multicast application potential receivers must be informed by the originator about the logical channel number being used.

This can be done either by non-broadcast means or, if prior agreement has been made, a well-known broadcast channel can be used for this purpose.

Close-down both of the connections between clients and access points and of the logical broadcast channels is by time-out. The time-outs for both connections and logical channels are long; well-known channels are never timed out. Clients wishing to receive permanently on a well-known channel must ensure that their receive route is not timed out if there is no activity on the channel for a long period. This is achieved by allowing reception clients to send a prompt to the access point. This prompt is simply reflected down the reception path but serves to reset the time-out. The prompt can be made distinguishable from normal data and is discarded upon reception.

3.6 Interconnection with other networks

For any new network architecture it is relevant to examine the possibilities for interconnection with the well-established architectures that have been widely implemented over the past decade or so. Two important types of network to consider are those that use datagram transmission as the prime means of communication between hosts and those that offer traditional, heavyweight virtual circuits. During the course of the project, interconnection was made to networks of both these types.

Since the UNIVERSE network architecture offers a datagram service, the task of linking to another datagram network is, in principle, quite straightforward. Extensive study of the problem of concatenating computer networks has been carried out by the DARPA community and this work has resulted in the specification of an internet datagram protocol (IP) (RFC791 in SRI, 1982*a*), which has been widely implemented. One means of interconnection with networks using the internet protocol would be to encapsulate the IP datagrams within UNIVERSE datagrams, with a linking host performing the operations of adding and removing the UNIVERSE datagram headers. Receiving packets from the internet, this linking host would have to perform an internet to UNIVERSE address translation; it would also need to fragment long IP datagrams (or datagram fragments) in order to convey them within UNIVERSE datagrams. This approach provides a simple means to enable the UNIVERSE network to be a subnetwork of the internet. Such a scheme implies that UNIVERSE hosts have to be programmed to deal with IP as well as the UNIVERSE datagram protocol.

A more restrictive method of connection to a datagram network (Cole *et al.*, 1983), which gave partial transparency, was designed and implemented during the project to interconnect the UNIVERSE network with the STELLA network (Celandroni *et al.*, 1983), which used a form of internet datagram. The UNIVERSE network was treated as a subnetwork in the STELLA catenet. Hosts on UNIVERSE participating as STELLA internet hosts implemented the STELLA internet protocol and encapsulated the datagrams inside UNIVERSE datagrams for transmission across the UNIVERSE subnetwork to other parts of the STELLA catenet. The STELLA internet relays built into the STELLA LDCs (see Chapter 4) performed a fixed address mapping between the two subnetworks (though this was supplemented by the possibility of treating STELLA addresses as logical addresses and subjecting them to a further level of indirection by look-up in a table). The circumstances of this interconnection were rather special as the two networks shared use of the same satellite channel. In effect, the satellite bridges became the linking hosts and the obvious optimization was incorporated to avoid double hops through the satellite.

One of the most widely implemented heavyweight virtual circuit protocols is that specified by the CCITT in its X.25 recommendation (CCITT, 1981*a*). No direct network level interconnection with such a network is possible since the nature of the native network protocols and services offered by each are different. In current ISO OSI terminology (ISO, 1986), a *relay* must be used, with suitable *subnetwork convergence protocols* (SNCPs) in each subnetwork. (Distinctions between subnetwork dependent and independent convergence protocols (SNDCPs and SNICPs) are ignored in the simple model presented here.) A diagram showing the relationships in conventional OSI form is given in Fig. 3.8. The particular instance of this type of network with which the UNIVERSE network was interconnected was the UK Joint Academic Network, JANET. This network uses a form of X.25 protocol compatible with the UK Packet Switched Service (PSS) provided by British Telecom. Relaying between different component subnetworks is performed at the level of the UK interim transport service defined in PSS User Forum, 1980 (the UK 'Yellow Book'). (Since that definition was made, it has come to be regarded as a possible sublevel within the OSI network layer.) In the context of JANET, the component which performs this relaying is commonly referred to as a *gateway*.

As shown in Fig. 3.8, the UK 'Yellow Book' network service in JANET may be regarded as being realized by an SNCP defined by Annex I of PSS User Forum, 1980. This resides above a *subnetwork access protocol* (SNACP) consisting of X.25 level 3, which in turn rests upon a data link protocol (DLP) using HDLC (X.25 level 2). The corresponding arrangement on the UNIVERSE network side is an SNCP consisting of a minimal implementation of CR82 (Larmouth, 1982). CR82 is a standardized derivative of BSP, devised for use in connecting single rings to the JANET network; it differs from BSP primarily in the provision of fields in the OPEN and OPENACK containing the more general form of addresses recommended in PSS User Forum, 1980. This SNCP rests upon a SNACP consisting of the UNIVERSE LVC, which in turn is built above a DLP using basic block

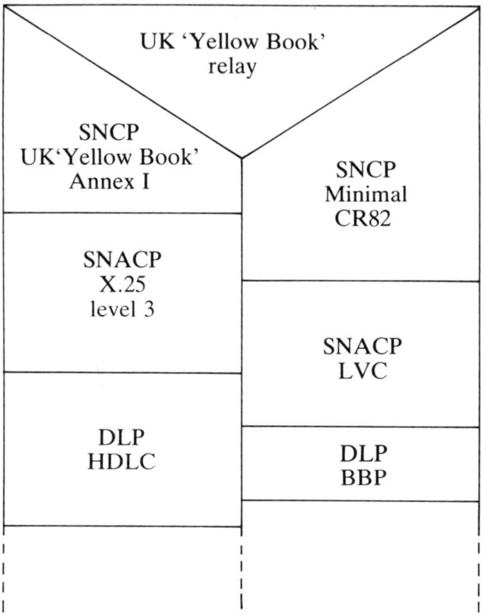

FIG. 3.8 JANET–UNIVERSE relay.

protocol. Details of lower layers (including media access control (MAC)) are not discussed here, particularly since UNIVERSE used more than one.

Gateways performing this form of relaying between the UNIVERSE network and JANET were built: details of their construction are given in Chapter 5. The most useful of their functions was to allow the use of one of the UK interim standard terminal protocols in use on JANET, TS29 (PSS User Forum, 1981), for remote terminal login from one network to the other or even to enable terminal communication between two UNIVERSE LANs on different sites to be accomplished by means of JANET when the satellite segment was not available.

These gateways were also used by a special type of bridge, called a *basic block tunnel* (Chapter 5), to provide an interconnection service between sites as an alternative to the satellite. The service was one which was equivalent, from the viewpoint of a host, in all respects but performance, to that provided by the satellite bridge. Here the heavyweight virtual circuit connections across the JANET network were used essentially as level 1 links to carry level 2 basic blocks between the sites. This facility was convenient for test operation in the absence of the satellite channel. It is hardly necessary to report that the performance of this arrangement was not spectacular.

3.7 Commentary

3.7.1 *Lightweight virtual circuits*

The driving force behind the UNIVERSE network architecture was the provision of LVCs as the primary service offered for communication between hosts. The choice of the LVC, as opposed to the reliable, flow-controlled, heavyweight virtual circuit, arose from a number of considerations; chief amongst these was the thought that burdening the bridges with the task of implementing error recovery and flow control would be a serious obstacle to the achievement of high performance. Another possibility would have been for the network to have offered only a datagram service. The use of datagrams also imposes a heavy load on bridges, as full address processing and route resolution must be carried out on every packet. In addition, datagrams provide neither a natural mechanism for blocking the initiation of new traffic-generating activities when the network is congested nor a means of guaranteeing bandwidth to particular activities. Nevertheless, this choice between LVCs and datagrams was not so clear and a datagram service was included in the architecture.

Error recovery in computer networks is normally realized by the use of an *automatic repeat request* (ARQ) scheme. An ARQ scheme is used in conjunction with an error-detecting code, such as a CRC, and block sequence numbers: when loss or corruption of data is detected by the receiver, the receiver requests, via a return channel, retransmission of the bad data. To be able to respond to retransmission requests, the transmitter must store data until a positive acknowledgment is received indicating that the data has been received correctly. If a negative acknowledgment is received, indicating loss of data, the normal recovery mechanism is for the transmitter to back up to the block in error and to retransmit that block and the blocks that follow. Clearly an important performance factor in any ARQ scheme is the round-trip delay, that is the time from transmission of the data to receipt of the acknowledgment. Implementations of ARQ are quite complicated since account must be taken of such circumstances as a requirement for repeated retransmission and the loss or corruption of acknowledgments.

In a network based on use of heavyweight virtual circuits, ARQ is usually carried out on each hop on the circuit, including the first and last hops which involve the communicating hosts. In an architecture based on LVCs, any ARQ is on an end-to-end basis and only involves the communicating hosts. The motivation for offering error recovery on a hop-by-hop basis is to reduce delays in acknowledgment; the end-to-end delay is broken up into a number of hop-by-hop delays and retransmission, if necessary, can take place after each of these shorter delays. This is clearly an important consideration when error rates are high, but has the implication that network components have the substantial additional task of ARQ to perform.

In circumstances where error rates are very low, as is the case for LANs, and high performance is required from network components, the alternative strategy of carrying out error recovery on an end-to-end basis seems preferable. In the case of a hop via satellite, an extremely long round-trip delay is combined with network components, namely satellite bridges, having a very heavy workload (see Chapter 4). In these circumstances the end-to-end recovery strategy seems to be particularly suitable.

Using LVCs, hosts may choose error recovery strategies appropriate to particular applications since these strategies are no longer imposed by the network. In cases where the chance of error is low, hosts may implement simple schemes, such as repeating whole operations, rather than attempting to correct corrupted data. The use of an ARQ scheme when real-time data such as voice and video are being transmitted is counter-productive since unpredictable extra delay is caused for some data. When error rates are reasonably low, the correct strategy for this type of data will normally be to leave errors uncorrected and rely on the inherent redundancy in the data (see Chapter 8).

Flow control on virtual circuits is the mechanism by which a receiver controls a transmitter to prevent data arriving at a rate faster than the receiver can handle. The procedure normally employed is related to that used in ARQ. The receiver acknowledges receipt of packets by sending messages on a return channel. The transmitter has a limited *window*, that is the number of packets it may have transmitted but which are outstanding awaiting acknowledgment. By introducing some delay before acknowledging receipt of packets, the receiver may slow the rate of transmission of packets on the circuit. As for ARQ, this type of flow control mechanism may be exercised on a hop-by-hop basis or end-to-end. Even when end-to-end flow control is used, some network architectures impose on clients the use of a specific flow control scheme because the mechanism will also serve to limit congestion in the network. When the store-and-forward components in a packet network become congested, that is when queue lengths within these devices increase, packet transit delays across the network increase. If such an end-to-end flow control scheme is mandatory, this extra delay will automatically have the effect of reducing the rate of flow of packets on virtual circuits and thus tend to reduce congestion.

Uniform imposition of this form of flow control is inappropriate for networks offering integrated services. In particular, on circuits carrying real-time traffic such as voice, methods of flow control that increase delay are unacceptable. A method of control that is acceptable is to limit the maximum rate of flow associated with a virtual circuit at the time the circuit is established. However, having allowed the initiation of the circuit, the network should be able to sustain this maximum rate of flow when required.

The implications of this requirement are that the network should be able to restrict the number of virtual calls opened, that is to block the initiation of new calls, and that the network should have a means of giving some traffic priority. Devising a priority scheme for datagrams is quite straightforward, though it does not guarantee bandwidth to an application; however, there is no natural means for blocking 'calls' when datagrams are used. Neither call blocking nor guaranteed bandwidth service was implemented on the UNIVERSE network, though these would be essential features for a network offering integrated services via the use of LVCs.

A feature of the architecture was the decision not to provide any explicit close primitive in the network service and instead to allow all LVCs and return paths generated by SSPs to time out. Principal among the arguments in favour of this scheme were those of simplicity and compatibility with single ring operation. Timing out is an appropriate mechanism for closing paths created to serve transactions and for these a short time-out may be applied. Moreover, some form of time-out closure must always be implemented to safeguard the network against the consequences of clients failing to close circuits after use. The provision of an explicit, client-initiated close operation in the architecture would need either the definition of a special packet that could be recognized by bridges or the addition of a new, out-of-band mechanism. Neither of these schemes would be compatible with single ring procedures and adoption of the in-band solution would make it necessary for bridges to examine every passing data block. It should be noted that use of time-outs does require extra resources in the form of the use and management of a large range of port numbers, particularly as it is prudent to leave these resources dormant for a significant period after use. In practice, the time-out mechanism was found to be quite satisfactory in the project, although there was generally little testing of the mechanism with the network heavily loaded by traffic generated from a large number of circuits.

3.7.2 *Transactions*

To keep compatibility with the services used by client systems on a single ring, a transaction service, the single shot protocol, was designed explicity into the network architecture. The SSP was implemented essentially as the call connection phase of an LVC, with no forward path set-up and with the return path having a short time-out. Although very convenient for client systems, the processing load per transaction placed on the bridges by this method of implementation is inevitably rather large, comparable with using datagrams. This load on the network components can be largely removed if an LVC connection is opened when a series of transactions are being carried out between a pair of hosts. The transactions may then be carried out across this connection, either using a simple 'request' and 'reply' procedure as in SSP or a more structured form of transaction, such as might be provided by *remote procedure call* (RPC) style protocols (Hamilton, 1984; Nelson, 1981), which deal with questions of reliability, type checking of data structures, and synchronization. Isolated transactions between hosts may still best be handled either by the 'truncated call' method of the UNIVERSE architecture or by using datagrams as the carriers. Transactions were heavily used in the experimental applications carried on the network; the SSP protocol, acting as a primitive form of RPC mechanism, was found to be quite effective in this context.

The problem of calculating suitable time-out periods to associate with transactions, for either SSP interactions or LVC establishment, is quite difficult. The delay between request and response is made up of two parts, the network transit delay and the processing delay in the server. The UNIVERSE nameservers gave some information helpful to clients, in the form of two single-bit flags indicating whether the service was located on the local ring or on some other ring and whether a short or long processing delay was to be expected. This information is insufficient, particularly since the only recourse available to a client after a failure is to retry.

When the path to a destination is known to the nameserver, an order of magnitude estimate of the expected network transit delay should be given which takes account of the number of hops required and whether any of them are slow. When this information is not available, this should be indicated to the client and a minimum expected delay given. Indications of expected processing delays given by the nameserver are often of limited use; as well as queuing delays, further nested transactions may be involved. A more satisfactory solution in these circumstances is for the server to give some holding reply to the initial request, to indicate that progress is being made. This problem is not peculiar to the UNIVERSE network but is emphasized by the incorporation of a transaction mechanism into the network access protocols.

3.7.3 *Nameservers and topology*

Nameservers were built in as an integral part of the network architecture, with the primary purpose of allowing clients to use names rather than addresses for accessing services. These servers, however, also played an important part in the call establishment mechanism of the network, the objective being to give users the possibility of dealing with local and remote services in a uniform manner. This objective was realized but led to the use of an unsymmetric procedure for call setup, in which the call request was signalled out of band, that is not using the path of the virtual circuit being set up, but the acknowledgment indicating the forward path to be used was received by the call requester in-band on the virtual circuit. This reflected the procedures pertaining to a single ring where call establishment proceeded by direct negotiation between hosts, without the nameserver being involved. This unsymmetric procedure is unnecessarily complex and, using a 'call setup' server, it could be simplified and made fully out of band. Such a 'call setup' server would make use of directory-type services on behalf of its client.

The only directory service used in the project was an elementary mapping of names to addresses. Other information which in general might be required includes routing tables; the fixed routing scheme used in the project was implicit in name-to-address resolution (see below). Provision of such services as part of the network service implies that supporting database management facilities are made available. Essentially this is a rather specialized problem in distributed database techniques, with requirements for easy, secure update and rapid consistency checks. Provision of such facilities for a large network is not simple and was not undertaken by the project.

In the project implementation of the architecture, partial information about the topology of the network was kept in both the nameservers and the bridges. Each nameserver knew which bridge on its LAN should be used to reach the (fixed) single exit point from its local site and which bridge led to each of the other LANs on the site. Each bridge held the corresponding information for both the LANs that it joined; the exception to this was the site exit point, the satellite bridge, which only had to hold this information for the single LAN to which it was connected. (Note that the ring at Logica was technically a subnet of the UCL site.) This arrangement worked adequately in the project since the topologies within sites were fixed and relatively simple and all inter-site connections were single hop via the satellite. In larger networks where both site topologies and inter-site routing may be quite complex, this arrangement would not be satisfactory. In general, bridges should only have to deal with simple tables which are updated infrequently. Complex routing decisions should be undertaken by a server that is not involved in continuous high-performance activities and that has access to database facilities. There still remains the question of how such a server might convey specific routing information to the bridges along a route. Although some solutions, such as source route addressing, are certainly known, no investigation of this question was made during the project.

3.7.4 *Conclusion*

A fundamental objective of the project was to provide an architecture which linked LANs together in such a way as to provide a wide area network with behaviour and services like those of a single LAN. This, in general terms, was clearly achieved. At a more technical level, a central objective of the architecture was to provide a set of network access protocols and services compatible with those in use on a single LAN, specifically the ring-based distributed system at Cambridge. This also was largely achieved. The LVC and SSP specifications and behaviour were sufficiently compatible with those of the single ring that in many cases application programs worked without change over the UNIVERSE network. Moreover, the network access protocols were found easy to use even by those inexperienced in the intricacies of such protocols.

A novel feature of the architecture was the introduction of a set of network access broadcast protocols. The original incentive to develop such protocols arose from the intrinsic broadcast nature of the satellite channel, coupled with its expense and a corresponding desire to explore use of the broadcast feature to optimize bulk data transfer. Some work in the application of broadcast has occurred in the context of distributed systems on LANs, especially for system startup announcement and resource location; relatively little has been done on wide area networks, largely because of a lack of support for broadcast. The architecture of the network access broadcast protocols proposed here seems interesting from two points of view. Cursory consideration suggests that it could support known applications, such as those just mentioned. Furthermore, the definition seems to offer some independence from the details of the physical mechanisms available in underlying segments of the network; the UNIVERSE network itself served as an example of this, in that the space segment was intrinsically broadcast but the LAN segments had no low-level broadcast mechanism.

A wider objective of the project was the investigation of the behaviour of such an architecture and its suitability for carrying a wide range of traffic. To this end, a substantial number of applications were developed and tested during the operational period of the network of a little over a year. These are described in later chapters. The applications covered a very full range of activities and made heavy use of both the LVC and SSP services. A year of activity on a smallish network is only sufficient to give some preliminary indication of the strengths and weaknesses of a network architecture. However, the results and experience obtained were sufficiently encouraging to suggest that further evaluations of this type of architecture should be undertaken.

4. Satellite access

4.1 Introduction

The UNIVERSE network made use of a broadcast satellite communications channel to give wide area connectivity between the sites making up the network. The satellite channel provided an overall transmission capacity of 2 Mbit/s between the sites. This capacity was reduced to 1 Mbit/s for data by the use of half-rate encoding for forward error correction, improving the bit-error rate for the channel to 1 part in 10^9. Thus the satellite portion of the network had similar capabilities in speed and low error rate to the LANs used for intra-site communication and allowed such applications as voice and slow-scan TV transmission as well as the use of distributed computing techniques, together with the more traditional computer networking activities of file transfer and remote login.

A key component of the UNIVERSE network was the satellite bridge which linked the LANs at a site to the satellite channel. In a fashion similar to stations on a LAN, these bridges act as stations on a satellite network providing access control to the satellite channel. The bridges were in fact the only stations on the satellite network. All traffic between bridges was sent in the form of addressed packets, with bridges receiving all packets but accepting only those with their own address or with the *broadcast address* (cf. Section 4.2), in the same manner as stations on a LAN. To share access to the satellite channel for transmission a time allocation scheme was used, making it necessary for the bridges to synchronize their timings and to control their transmissions to stay within their allocated time slots. Both the transmission of the signals required to keep the timings of the bridges in synchronism and the control of time allocations to individual stations were carried out centrally by a *master* station. This master station, however, also functioned as a normal station and it was possible for any station to take on this master role as all the bridges were of identical construction and ran the same software. The time allocation scheme used in the project reflected the requirement for bridges to be able to carry traffic generated by varying applications with quite different characteristics. The scheme employed provided time allocations for an individual site based on its own view of its total needs at any time. Each bridge calculated its transmission requirement at regular intervals and sent requests for time allocation to the master station based on this requirement. Using this information the master station allocated the available capacity and informed all stations of their time allocations in messages broadcast at regular intervals. Details of the protocols used to control access on the satellite are described in Section 4.2.

In terms of both hardware and software, the bridge was a sizeable and complex item of equipment. To save time on its construction, equipment that had previously been used in the European satellite communications experiment STELLA (Hine, 1984) was adapted for the purpose. This equipment used a minicomputer to drive a small satellite ground station via a microprocessor-controlled interface. The major bridge development work required for the project was the design and production of new software for the minicomputer and the earth station interface. The hardware development of a Cambridge ring interface for the minicomputer was also required, together with some modifications of the earth station interface. The software development for the satellite bridge was the largest single task undertaken in the project. Development of the bridge was started in mid-1981 and testing of the first bridge was complete by mid-1982. Commissioning of the bridges at all the sites occupied most of the second half of 1982. Details of the satellite bridge implementation are described in Section 4.3.

During the project a number of measurements were made of the performance of the satellite bridge and these are described in Section 4.4. Due to a number of limitations in the hardware of the bridge, particularly the built-in, half-rate Viterbi encoding and the long guard times between transmissions from different stations, the theoretical maximum data throughput of the satellite channel was limited to about 630 kbit/s rather than the 2 Mbit/s of the basic channel. Measurements also showed that the throughput of the bridge was processor limited in terms of the number of packets transmitted or received per second, with these numbers being essentially independent of packet size. The bridge software could sustain a transmission rate to the satellite of a little over 40 packet/s. Its performance receiving packets from the satellite was lower, being in the range of 20–30 packet/s. This difference was mainly due to the overhead involved in allowing the bridge to retransmit packets on the ring if the receiver failed to take the packet at the first attempt. No such retransmission was allowed on the satellite links. These rates are low but are not untypical of the performance of a minicomputer in this type of application. Using large block sizes of 1000–2000 bytes, the throughput in terms of bits per second is quite respectable and the bridge was able fully to load the satellite channel.

Use of existing equipment with known properties was dictated to the project by the very tight timescales. The shortcomings of this approach, particularly the performance weaknesses and cost of using a minicomputer, were known at the start of the project and accordingly a subproject was set up, within the main project, to design and construct a new purpose-built high-performance satellite bridge. A specification for this new bridge, including access controller, encoding scheme, and modem was developed and taken to a stage at the end of the project where a prototype was undergoing first tests. This work has been continued after the end of project UNIVERSE and is described in Section 4.5.

4.2 Satellite access protocol

4.2.1 *TDMA schemes*

The satellite access control mechanism used in project UNIVERSE was a form of time division multiple access (TDMA) scheme called demand access TDMA. Overall development of this scheme was carried out in collaboration with members of the STELLA project. Conventional TDMA schemes are used to subdivide or multiplex the use of a communications channel to provide a number of subchannels that in general may link different end-points. In these schemes a fixed-length time frame is defined and this time frame is subdivided into a number of slots. Time frames, which are repeated continuously, are normally introduced by some reference signal on the channel from which stations synchronize their own timings. Generally, each time slot within the time frame is allocated for the transmission of data between particular stations with one station transmitting and the other receiving. In some TDMA schemes slots may be assigned for broadcast communication with a particular station transmitting and some or all of the other stations receiving. The assignment of slots is normally for long periods (i.e. a large number of time frames), either for fixed time intervals or for the duration of a 'call', and is organized by a central administrative station assigned with this task. All stations must have permanent access to this administrative station.

In demand access TDMA, stations transmit data in the form of addressed packets and thus the requirement to associate time slots with destinations is removed. Time slots are now allocated to stations for the transmission of packets which may be addressed to any station on the network and stations may transmit data to several different destinations within the same time slot. The allocation of time slots, their size, and duration are varied with time to follow the demand that stations make for transmission capacity. As with different LAN technologies there are many different ways in which this allocation, or division, of the channel capacity can be carried out. However, for the satellite network, all methods have to cope with the long transmission delay. On a LAN, all stations can determine the use of the channel by another station within a time that is short compared with the time for the transmission of a packet of average length. On a satellite network, the dissemination of this information takes many packet transmission times. Schemes proposed for access control of packet transmission on satellite networks make use either of contention or reservation or a combination of these techniques. In contention schemes, stations transmit packets as soon as they are ready, irrespective of the activities of the other stations on the network. Collisions between packets transmitted by different stations are possible and may be detected by stations receiving their own transmissions relayed by the satellite. The problem with contention schemes is to achieve reasonable levels of efficiency in the use of channel capacity. Carrier sense schemes, such as CSMA (carrier-sense multiple access) that are used to give high channel efficiency in LANs using contention techniques cannot be used with satellites. Contention schemes allowing totally random transmissions without any synchronization between stations may be used but performance can be improved by a factor of two if transmission is synchronized into slots and fixed length packets used; even so, channel efficiencies are limited to about 30 per cent in favourable circumstances.

In reservation schemes stations transmit a relatively small number of bits to request channel time for queued data, and then subsequently transmit this data in allocated time slots. The problem with this type of scheme is to maintain responsive, flexible sharing of the channel in the face of the long transmission times for allocation requests. Three features characterize reservation schemes:

(1) the way in which reservation requests are sent;

(2) the allocation algorithm; and

(3) whether reservation control is centralized or distributed.

Stations that already have time allocated can use this to transmit new reservation requests, but some special provision has to be made for stations that are inactive and wish to start transmitting. A number of small slots are usually reserved within the time-frame for this purpose. These *reservation request slots* can either be assigned in a fixed fashion to stations or can be accessed by contention. A fixed assignment was used in the project. There are a number of different allocation algorithms that can be used such as first come, first served; round-robin; or priority oriented. The reservation mechanisms that have been studied most extensively, both theoretically and experimentally, are the priority-oriented demand assignment (PODA) schemes. In these schemes reservations may be made either for streams of data or for individual datagrams. Once set up, a stream gives automatic scheduling for packets to be transmitted at predetermined intervals until the stream is closed. This type of reservation is suitable for telephony traffic. Separate reservations must be made for each datagram transmitted. Two variations of the scheme called FPODA (fixed PODA), in which the reservation request slots have fixed assignments, and CPODA (contention PODA), where contention is used for the access to the reservation request slots, have been used. The scheme used in the project was a variation of FPODA in which stations based their requests for time allocations on their overall requirement for traffic capacity.

Reservation can be carried out by using either centralized control or distributed control. In centralized control schemes, reservation requests are sent to a master control station, which implements the reservation algorithm and broadcasts an allocation schedule to all stations. In distributed control schemes, all stations receive reservation requests and implement the reservation algorithm and are thus able to calculate the allocation schedule for themselves. Distributed control reduces the delay between request and allocation by at least one satellite round-trip delay time, but recovery from error, that is failure of a station to receive a reservation request correctly, is more complicated. Centralized control was used in the project, but all reservation requests were transmitted as broadcast, that is to all stations rather than a particular destination station. It would have been quite straightforward to have tested the reservation scheme with distributed control, but insufficient time was available.

4.2.2 *Details of protocol*

The format of the time frame used in the project is shown in Fig. 4.1. A master station is defined which, in addition to its functions as a normal station, is responsible for transmitting a reference signal at regular intervals to define the time frame. This station also has the task of controlling the allocation of transmission time to different stations. The time frame consists of the reference slot followed by a series of small slots, allocated permanently, one to each station on the network. The remainder of the frame is allocated on a demand access basis by the master station using variable size time slots. It will be seen that between transmission slots there are guard times during which no transmission takes place. These are imposed to prevent any jitter in transmission timings between different stations causing collisions of signals at the satellite.

All information, including the reference signal, is transmitted in packets using the HDLC packet format. This format provides a scheme of framing flags which enables stations to detect the start and end of a packet and a cyclic redundancy check (CRC) to allow stations to check that the information in a packet has been correctly received. Within the HDLC format, all packets start with a four-byte sequence followed by a variable-length data field. Hardware in the station uses the DESTINATION field

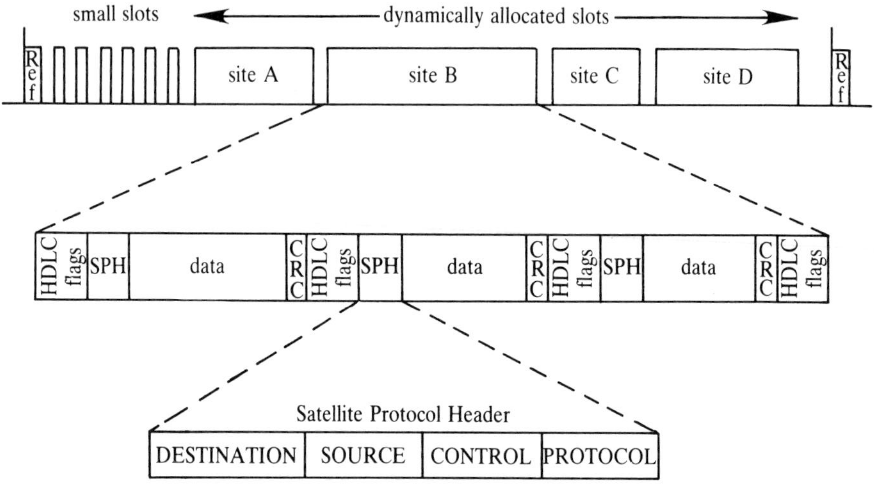

FIG. 4.1 **Format of time frame.**

to detect packets which are destined for its own station or are broadcast packets. Broadcast packets have the DESTINATION field set to 'all ones', the *broadcast address.* The SOURCE field is set to be the station which transmits the packet, except in the special case of the packet transmitted by the master station in the reference slot. This broadcast packet is in addition given 'an all' ones source field so that it is distinguished from all other packets.

It should be emphasized that, although packets were transmitted over the satellite using the HDLC packet format, the HDLC retransmission procedures were not used. In fact no retransmission after error was implemented within the UNIVERSE network; the error rate end-to-end across the network was very low and responsibility for any necessary correction of errors was assigned to the communicating hosts, not to the network service.

Two types of packet are carried on the network, packets carrying user data and packets carrying control information. The CONTROL field is set to all zeros for packets carrying user data and, for control packets, to a value indicating the control function associated with the contents of the data field.

Control packets are transmitted by stations to indicate their current status and to request changes in their transmission time allocations. Control information is sent by stations to the master station, but is sent with the broadcast DESTINATION address. This allows any station to act as the master station; stations not acting as the master station normally ignore control packets apart from the reference packet.

Finally, the PROTOCOL byte is available to distinguish between possible protocols that may be contained in the data field. This field was not used in the project and was kept constant.

In a single time frame, the reference slot and the short slots can only be used to carry single packets; the larger slots can each carry a number of packets. The purpose of the small slots is to give each station the possibility of transmitting control packets at all times, particularly on occasions when the station has not been allocated a slot in the dynamically assigned part of the frame. Thus, on startup, a station can announce its presence on the network and ask for transmission capacity using its small time slot. This slot may also be used to carry small data packets such as those generated to satisfy high-level protocol requirements. The master station transmits a control packet in every reference slot and uses this to inform all stations of the current slot allocations and the status of the network. This reference packet thus serves a dual purpose, to define the start of frame and to carry control information from the master station to all active stations on the network.

Two types of control message are provided, station status messages and capacity request and assignment messages. Stations are expected to announce or confirm their presence on the network by transmitting a 'hallo' message at regular intervals. If the master station fails to receive a 'hallo' message from a station within a fixed period, it will consider that station no longer to be active on the network and will inform all stations of this change of state using a control packet transmitted in the reference slot. Control messages are available to allow a station to close down gracefully and to allow the master station to close down. The latter message is again transmitted in the reference slot and indicates that another station should take on the role of master station.

For making capacity requests, three types of control message are available to stations: an assignment request, a relinquish request, and a traffic indication. In an assignment request, a station asks for a slot to be allocated to it, quoting a slot identification number, the quality of service required, and the minimum useful slot size. The slot identification number, together with the station number, serves to identify a particular slot and allows a station to be allocated more than one slot. Three types of quality of service are available, *priority*, *traffic indicated*, and *bulk data*. *Priority* is intended for services that cannot be pre-empted, such as telephone conversations, and a priority slot once allocated by the master station will remain in use at a fixed size until it is relinquished by its owner station. Using the second category, *traffic indicated*, a station periodically transmits a control packet giving the current length of its queue of packets awaiting transmission and this information is used by the master station to adjust the sizes of slots allocated to competing stations. ***Bulk data*** indicates that the master station should allocate any spare capacity to this activity. Bulk data has the lowest priority. The master station receives allocation requests and traffic indications and uses this information to share the transmission capacity between the stations wishing to transmit and to adjust the arrangement of time slots in the dynamic area of the time frame. It periodically broadcasts new arrangements of time slots in the reference packet, giving for each slot the station number and slot identifier together with the offset times of the beginning and end of the slot measured relative to the beginning of the frame. The master station in its role as an ordinary station also sends its allocation requests via the satellite to itself as master, ensuring that all stations obtain equal treatment for their requests.

Finally, a protocol for dealing with broadcast data was defined. There is no problem in transmitting data in a broadcast mode as by setting the DESTINATION field at the start of each packet to the 'all ones' broadcast address, the transmitter will cause all stations to receive the packet. Receiving stations can also easily distinguish data from control messages by examining the contents of the CONTROL field. To allow the network to carry a number of broadcast applications simultaneously, the port number field of the basic blocks being carried as data was used to define broadcast channels. Since some applications require that several stations transmit on the same broadcast channel, allocation of these port numbers has to be carried out centrally. Control messages to support this central administration of broadcast port numbers by the master station are defined in the protocol. As far as the satellite network is concerned, these are the only protocol extensions that are required for broadcast data transmissions.

4.3 Satellite bridge implementation

4.3.1 *Components*

The hardware of the satellite bridge consisted of a satellite earth station with associated modem and codec, a minicomputer and two microprocessor-driven interfaces, one for the earth station and one for the connection to the Cambridge ring. The arrangement of the hardware is shown in Fig. 4.2. The earth station used was a standard, small-dish, Ku-band station supplied by Marconi Communications Systems Ltd. Ku-band stations are defined to transmit in the 14 GHz band and receive in the 11.5 GHz band. The station consisted of a fixed, 3 m diameter, roof-mountable antenna with wave guides leading to a six-foot rack of electronic equipment, which included the transmitter amplifier, the modem, and the codec. The high-power amplifier used a travelling wave tube (TWT or klystron) and gave the station a maximum transmission power of 140 W. In order to minimize the noise in

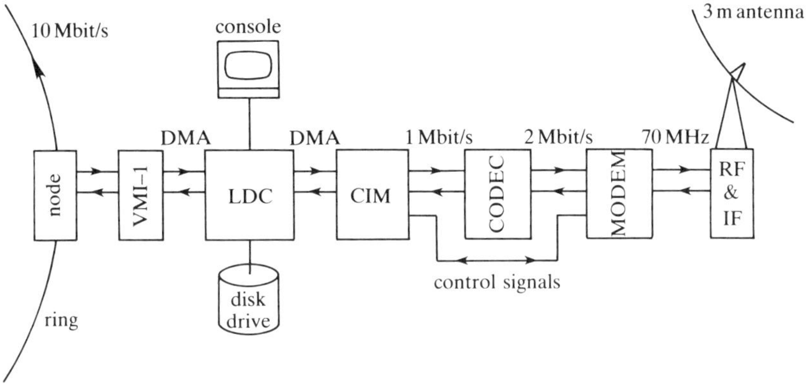

FIG. 4.2 Satellite bridge hardware configuration.

the signal reception path, it is advisable to amplify the signal as near to the physical point of reception as possible, using an amplifier with low noise properties. This is particularly necessary with a small antenna having correspondingly low gain. For this reason a field effect transistor amplifier was mounted on the back of the antenna dish. The conventional figure of merit used in (satellite) communications engineering is the ratio of system gain to equivalent noise temperature, G/T, since this is a measure of a channel's ability to convey uncorrupted information. G/T for the earth stations used in the project was 22.5 dB/K. The transmitter and receiver amplifiers were driven by a 2 Mbit/s burst BPSK modem which worked in a 5 MHz portion of the bandwidth covered by one of the transponders of the OTS satellite. Other applications, independent from project UNIVERSE, were able to use the same transponder working in other parts of its bandwidth. The modem in turn was driven by an half-rate codec using convolutional encoding and Viterbi decoding (Viterbi and Omura, 1979), supplied by Linkabit Corporation, USA. It was not possible to bypass the codec to send unencoded data. However, the combination of transmission power, receiver G/T, and the half-rate encoder produced a communications channel with a very low bit-error rate of the order of 1 bit error in 10^9 bits transmitted. Thus from the computer networking point of view this earth station equipment gave access to a wide-area broadcast communications channel with a data rate of 1 Mbit/s and a bit-error rate of 1 in 10^9.

The minicomputer, which for historical reasons was called the link driving computer (LDC), was a GEC 4065 with half a megabyte of memory. The LDC was connected by DMA to the earth station interface, which was controlled by a Motorola 6800 microprocessor. This interface, called the communications interface module (CIM), was designed and built by the STELLA group at the European Organization for Nuclear Research (CERN) Laboratory, Geneva. It drove the 1 Mbit/s transmit and receive serial links to the earth station. The main tasks of this interface were to organize the timing of transmissions to the satellite and to convert packets to the HDLC format by carrying out the bit-stuffing operation and providing the CRC processing. These latter operations were implemented in hardware. The control program for the CIM was loaded from the LDC. This software was developed by CERN and RAL in collaboration.

The LDC was interfaced to the ring by a proprietory microprocessor controlled interface board, called a Polynet VMI-1, originally made by Logica VTS Ltd. This board was connected by DMA to the LDC and driven by an Intel 8086 microprocessor with fixed software in read-only memory. Its main function was to support the basic block protocol on the ring.

4.3.2 *Bridge software*

The software in the LDC was structured as three processes running in parallel: a CIM driver, a ring driver, and the main bridge process. These processes executed under the GEC operating system OS4000. Communication between processes was by interprocess message, a facility of the operating system supported by the microcoded nucleus of the 4065. The bridge process was written in BCPL and the drivers were written in Babbage, the high-level assembly language of OS4000. The bridge had to carry out five main functions, as follows.

(1) Organizing the flow of packets from the ring to the satellite network and vice versa.

(2) Handling the UNIVERSE network protocols, OPEN (BRIDGEOPEN), OPENACK, SSPREQ (BRIDGESSPREQ), SSPRPLY, datagrams, and port mappings on established connections.

(3) Handling the satellite network protocol, described above in Section 4.2. In the case of the master station, this includes allocating time and broadcast channels.

(4) Supporting the ring basic block protocol.

(5) Organizing the buffer pool and queues necessary to support these functions.

In dealing with the satellite network protocol the bridge process was assisted by the CIM in all matters concerned with timing. The CIM would recognize reception of the reference packet and reset its timers relative to the start of the time frame. It would accept commands from the LDC to update the position of the station's transmission bursts relative to the start of the time frame and ensure that transmission from the site stayed within these timing windows. If the station was acting as the master station, the CIM would accept packets, normally containing time allocations, from the bridge process for transmission in the reference burst and would ensure that reference bursts were transmitted at regular repeated intervals, that is at the start of each time frame. The CIM contained HDLC chips which formatted the data into HDLC packets and filled inter-packet gaps with framing flags.

On the ring side the VMI-1 interface board dealt with the basic block protocol and the bridge process was only responsible for organizing the retransmission of blocks that were not received correctly.

4.3.3 *TDMA implementation*

The length of the TDMA time frame chosen for use in the project was 130 ms. This time, which causes delay in the transmission of packets, is rather long for such applications as real-time voice transmission but was necessary because the LDC was not powerful enough to split data packets across different time frames and at the same time maintain adequate performance. A basic block may contain up to 2048 bytes, which takes 16 ms to transmit and an additional 20 per cent must be allowed for possible bit-stuffing for the HDLC format. Thus 130 ms seemed a reasonable compromise between keeping the delay time low and still allowing room for a reasonable number of packets to be transmitted in a time frame.

It was necessary to compensate for the different distances between the earth stations and the satellite. Taking the satellite position to be fixed relative to the Earth's surface, the time delay T_i for a signal to travel from the ith earth station up to the satellite and back can be measured for all stations. For the UNIVERSE stations this round-trip delay varied from 260 ms for Loughborough, the most northerly station, to 254 ms for the demonstration site in Geneva. Consider the start of a reference packet leaving the satellite at time $t = 0$. This will arrive at the ith station at a time $t = T_i/2$. If this station starts to transmit a packet in a window offset by a time W from the start of frame, this packet will start to arrive at the satellite at a time $T_i + W$. In order to avoid signals from different stations colliding at the satellite, it is necessary to ensure that whichever station were to transmit into a window with offset W, the transmission would reach the satellite at the same time.

Consider the ith and jth stations and suppose that the ith station is further away from the satellite. Two possibilities immediately present themselves: either the ith station transmits early or the jth station delays its transmission. Since stations are required not to transmit in the current time frame unless they have received the current reference packet, the first of these possibilities is ruled out. For each station, a delay time, D_i, is defined. If the ith station is to transmit in the window with offset W, it actually transmits at time $W + D_i$ after the receipt of the reference burst, so that the transmission now arrives at the satellite at time $T_i + W + D_i$. Each D_i is chosen so that

$$T_i + D_i = \text{constant}.$$

This constant may be considered the value of the round-trip time for the site (real or hypothetical) with no associated delay time. This is the universal effective round-trip time used by the TDMA controller: it is chosen to be as small as possible, consistent with keeping all D_i non-negative.

In practice, the satellite is not stationary and each terminal should periodically re-measure its round-trip delay time T_i and adjust its delay time D_i. If T_i is found to have increased to $T_i + t_i$, D_i should be adjusted to $D_i - t_i$. This procedure is quite straightforward since stations can receive their own transmissions. This re-timing operation was not carried out in the project and the delay times used for each station were kept fixed. For performance reasons in the CIM (see Section 4.5) it was necessary to allow a guard time of 1 ms between transmission bursts from different stations and this was more than sufficient to allow for movement of the satellite.

It was always necessary for one of the bridges to take on the role of master station and generate the reference packet. Since all bridges contained identical software it was possible for any station to take on this role. To make operational procedures as simple as possible, the sequence of events at startup for a site was as follows. The site would first listen in non-master mode. If another site was transmitting the reference packet, it would stay in this mode and begin operation. If no reference packet had been detected in a period of 10 s, the site would take on the role of master station and begin transmitting the reference packet. The master station had to listen for control messages from other stations, in particular allocation requests, and implement an allocation algorithm. The allocation schedule would then be broadcast in the reference packet. A simple scheduling algorithm for the *traffic indicated* type of service was used during the experiment. In general the time available for allocation in each frame was divided equally between requesting sites providing their throughput indications remained below a certain threshold. Above this threshold a site was deemed to be 'backlogged'. In this case, other sites would be allocated the minimum amount of time to allow the transmission of one full size basic block, and the remaining time was divided equally between the 'backlogged' sites to give them larger time windows.

4.3.4 *Congestion*

In the same way as any other type of bridge it was necessary for the satellite bridge to implement congestion control. For the direction of flow from the ring to the satellite network, this control was quite straightforward; when the queue of packets waiting to be transmitted to the satellite reached a certain length, the bridge did not permit any further reception from the ring until the situation had cleared.

Congestion generated in the opposite direction, that is from the satellite to the ring, is much harder to deal with. In normal circumstances the ring is faster than the satellite network and the bridge should be able to forward data received from the satellite rapidly. However, if the ring is congested or if a receiver on the ring is busy and will not take data, queues may build up in the bridge. In this case it is not possible to shut off reception from the satellite quickly because the time required to inform other stations of the congestion is at least one satellite transmission delay time, that is 260 ms. Because virtual circuits are used in the UNIVERSE architecture it was possible for the bridge to use a scheme for discarding packets that caused minimum disruption, when this type of congestion occurred. In the circumstances where the queue of packets awaiting transmission to the ring grew too large, the bridge discarded packets belonging to the circuit (port mapping) which had most packets queued. This produced the most free space with the least disruption since only a single circuit was disturbed; moreover, on a given circuit, it usually makes little difference whether one or a number of packets are lost.

4.3.5 *Development and testing*

Clearly the testing and verification of a system such as the satellite bridge, which contains both new hardware and software, is a major task. Fortunately, a great deal of this testing can be carried out on a single bridge by looping back at various stages within the channel to the satellite, first within the CIM interface, then within the earth station, and finally at the satellite itself using the very useful property that a station receives its own transmissions reflected by the satellite. However, in the later stages with the verification of the TDMA scheme, tests have to be carried out using more than one bridge. In principle a number of bridges could be set up on a single site for this testing, but the expense and complication of moving and installing the equipment involved would be very high. In the project this testing of a number of bridges working together was greatly helped by the use of an X.25 packet-switched network.

The GEC 4065 minicomputer used as the LDC was equipped with a small disk and the bridge software executed under the control of the GEC

OS4000 operating system. This operating system contained X.25 networking software and the LDCs of the bridges at all sites of the network were connected to an X.25 service network, either the JANET network (Wells, 1984) or the BT packet network, PSS, these networks being connected together by a gateway. Three principal services were available on the X.25 networks, terminal sessions (remote login), file transfer, and electronic mail. These services were heavily used in both the development and the operation of the satellite bridge.

The source code of the bridge software for both the LDC and the CIM was kept on a GEC 4090 computer attached to the JANET network and located at RAL. This machine was used to compile the source code to create executable modules, which could be transferred as binary files across the X.25 network to the LDCs using standard network file transfer. These modules were stored on the disks of the LDCs and loaded for operation by console command. Testing was carried out by editing the source code and recompiling on the GEC 4090. Since the GEC 4090 was accessible from the X.25 network, this operation could be undertaken from any site. The code could be sent out and reloaded into the LDCs at the various sites under test in a matter of minutes. For operational testing of the software, remote login to the LDCs involved was used. By this means it was possible to drive a local and remote bridge using two terminals placed side by side. Parameters could be adjusted to allow various levels of monitoring to be shown on the consoles. Where extensive monitoring data was required for tests, this could be written to the local disk during operation and then shipped back for subsequent printing using file transfer. Since all transmissions from the satellite are broadcast, it was even possible at a local site to examine the timing characteristics of a remote station using a logic-state analyser attached to the receive side of the local station. Using these tools it was possible to avoid totally the very tedious and slow method of testing using technical experts at each site reporting information via long telephone calls. In one test five bridges were driven simultaneously from RAL using five terminals standing side by side. In retrospect, it is fairly clear that it would have been next to impossible to develop the satellite network in the time available had use of JANET not been possible for transferring new LDC software and providing interactive access for monitoring and debugging all the LDCs.

Problems were sometimes encountered when testing caused the bridge program to terminate abnormally. Usually, since the LDC was running under a standard operating system, the program could simply be reloaded by command from a remote terminal. Occasionally the problem might be sufficiently severe to require reconnection of the X.25 terminal call, and on very rare occasions it was necessary to resort to the telephone to ask someone at the remote site to reload the operating system of the LDC.

Construction of the hardware for the satellite bridges started in mid-1981. Work on the software was started at the beginning of 1982. Testing of the bridge systems occupied the second half of 1982 and the bridges were used operationally in the network during the year 1983.

4.4 Performance measurements

The aim of the measurement programme carried out on the satellite bridge was to find the capacity and limitations of the bridge as a specific component of the network. All the measurements were conducted using generated data and did not rely on any protocols of a higher level than the LVCs described in the previous chapter. Some measurements were taken during normal operation of the network, others used special versions of the bridge software which measured activity at one site whilst simulating activity from other sites. Although the TDMA controller would have been capable of providing the three qualities of service mentioned in Section 4.2.2, use of these was reserved for later experimentation with a mixture of types of traffic. In the event, experimental access to OTS ended before the project reached the stage of carrying out such experiments and in fact access to quality-of-service parameters was never provided in the network access protocols.

The maximum data bit rate through the satellite was 1 Mbit/s after encoding, assuming that the entire channel was being used continuously. The channel was split into divisions, first between stations and second between packets and each of these divisions imposed an overhead. For real data, use of the HDLC protocol imposed a further overhead from the bit-stuffing operation but artificial data was generated for measurement purposes which avoided this overhead. Out of the 130 ms time frame, the reference burst and the short bursts together with guard times between bursts took 40 ms. Assuming that all the remaining time was allocated to one site using maximum length packets of 2048 bytes, this site could transmit 5 packets per time frame giving a bit rate of $5 \times 8 \times 2048$ bits in 130 ms which is approximately 630 kbit/s. With several sites competing for time on the channel this maximum throughput rate dropped to about 500 kbit/s on account of the additional overhead.

4.4.1 *Round-trip delay measurements*

The round-trip delay was measured by sending successive blocks from a test machine round the ring to the bridge, up to the satellite and back down to the bridge, and round the ring. Figure 4.3 shows the average round-trip delay for various block sizes. The graph shows the delay for different block sizes in two circumstances, one with only one site active and the other with a simulation of three additional active sites. The graph does not show the delay imposed while the connection is set up; this delay, which occurs

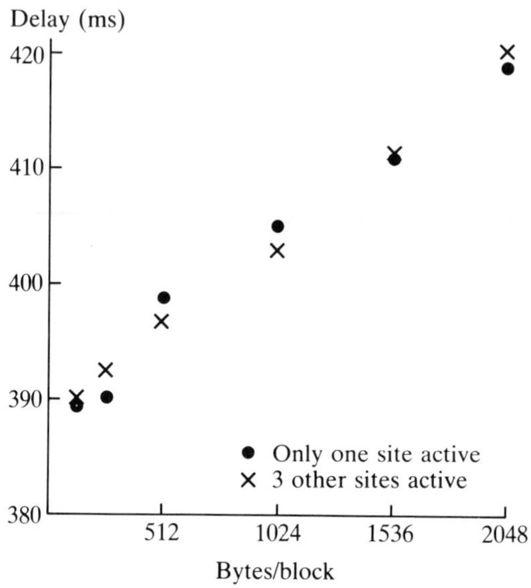

FIG. 4.3 Round-trip delay measurements.

while the station uses the satellite network protocol to negotiate a new transmission window, was between 1 and 1.5 s. Extrapolating backwards indicates that the measured delay for a block of length zero was of the order of 390 ms, which can be compared with a theoretical minimum average delay time made up of the propagation delay of 260 ms plus a delay of half the time frame, that is 65 ms, giving a total delay of 325 ms. The additional 65 ms represents the handling delay in the bridge and the bridge and station interfaces to the ring.

4.4.2 *Maximum transmission rates*

The maximum transmission rate that the bridge could sustain was measured by sending blocks from a station on the ring whenever the bridge was prepared to receive blocks, that is by keeping the bridge continuously active. The transmission rate was then calculated from the elapsed time between sending successive groups of 100 blocks. The blocks were sent to another inactive site, so that effectively the blocks disappeared once they had been transmitted by the bridge to the satellite. The results are shown in Fig. 4.4 as graphs of transmission rate against block size with different numbers of sites in operation, the activity of the other sites being simulated. For a bridge with no performance limitations, the graph would show a series of horizontal lines, representing the fraction of the total channel ca-

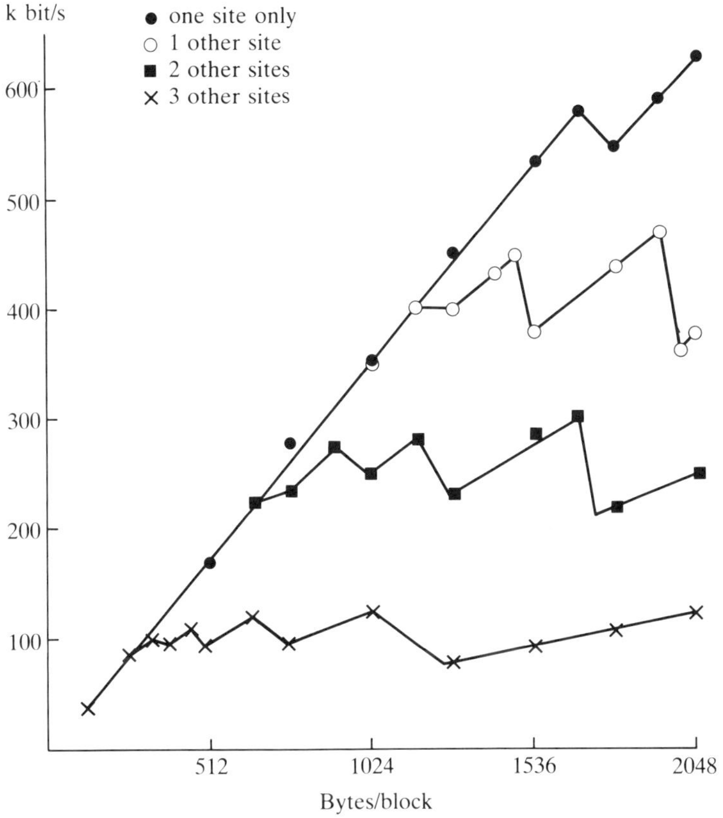

FIG. 4.4 Transmission rate vs. block size.

pacity allocated to the bridge. The jagged nature of the lines on the right of the figure corresponds to the limitation that the bridge could only transmit whole packets within its transmission window and could not divide packets so that parts of packets could be transmitted. Thus, in general, it was not possible for a bridge to fill its transmission window. The results show that a maximum rate of 630 kbit/s was possible when maximum block size packets were being transmitted. The fall away of the curves as the block size is decreased shows that the bridge is limited in the number of blocks per second that it can handle. The curves show this limit to be about 43 block/s irrespective of block size. The master station was responding to throughput indications for all the measurements shown in Fig. 4.4. The other sites were active but not 'backlogged' during the measurement, whilst

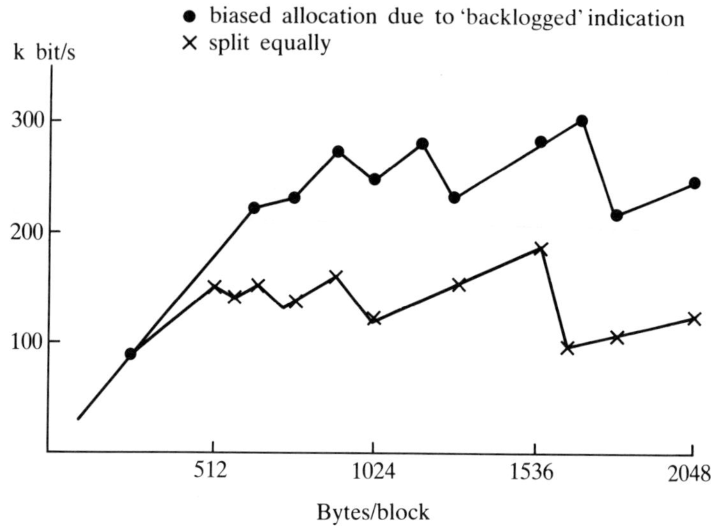

FIG. 4.5 Advantage given by throughput indication.

the site being measured was kept in a 'backlogged' state. Figure 4.5 shows the improvement that this technique gives over an algorithm that simply splits the time equally between active stations.

4.4.3 *Performance on reception*

Measurement of bridge performance on reception was a harder task. It was possible, using a special echo facility built into the bridge, to direct connection paths up to the satellite and back to the transmitting station, but use of the facility would have involved the bridge in both reception and transmission at the same time. The measurements were performed by using stations at two sites, one as a source and one as a sink. The remote station, which was used to generate traffic, had to be loaded and controlled across the network. The results of the first measurements were disappointing and showed that reception started to become unreliable at just below a rate of 20 block/s and at 25 block/s substantial numbers of blocks were being dropped. Again this maximum reception rate was essentially independent of block size. This poor performance was found to be due to the overheads in the bridge involved in retransmitting onto the ring blocks that had failed to be accepted by the destination station on the ring. Because of the requirement to maintain sequentiality for block delivery, the bridge could not chain blocks together for its ring interface to handle autonomously, even when these blocks were destined for a single destination. This was because, if the transmission of the first block in a chain were to fail and

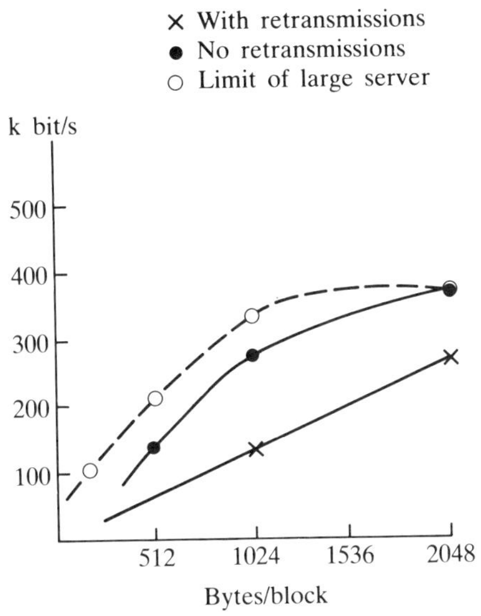

FIG. 4.6 Reception rate vs. block size.

a subsequent block transmission succeed, retransmission of the first block would result in blocks arriving out of order at the destination. This need to handle blocks singly for transmission onto the ring involved a substantial overhead of interruption and rescheduling in the LDC and in fact bridge reception performance when two different destinations were involved was much better. The retransmission facility was removed from the bridge and performance improved to a rate of 34 block/s. Figure 4.6 shows the measured reception bit rate against block size. The rate at maximum block size is a projected rate as the computer generating the measurement data could not keep up transmission at the required rate.

Even with the removal of the retransmission capability, the performance of the bridge for reception still did not match the transmit performance and this would have been a problem for serious investigation had it been proposed to continue with the use of this equipment.

4.4.4 *Other measurements*

Perhaps the most interesting measurement experiment carried out on the satellite network was the so-called 'chain-letter' experiment. In this experiment ring stations at each of a number of sites were arranged in a logical, circular chain, with the station at one of the sites acting as the source station. This station opens a connection to the second station in the chain and starts sending test data. This second station in turn opens a connection to the next station in the chain and relays the data it receives from the source station. The process is repeated along the chain with the last station relaying the data back to the source. By varying the rate of data transmission by the source the satellite channel can be loaded to a variety of levels with a number of different bridges contributing to the level. Using maximum size blocks of 2048 bytes the channel could be loaded by a chain of four stations to carry about 500 kbit/s, that is with each of the four stations transmitting one block in each time frame.

The 'chain-letter' scheme seems a potentially excellent way of producing a realistic controlled background load against which the performance of individual connections may be measured. That this possibility was not exploited during the project was due solely to lack of time.

4.5 A high-performance satellite access controller

In describing the TDMA system and its implementation used in the project, a number of weaknesses have been pointed out. It should be noted that most of these were shortcomings of implementation rather than in the TDMA design itself. During the project, a parallel development was undertaken to produce new controller hardware. Within the limits of the project as a whole and the equipment used, the behaviour of this type of TDMA scheme was considered sufficiently promising to justify further investigation, coupled with continued development of the new, higher-performance access controller. As well as rectifying many of the known weaknesses of the equipment used in UNIVERSE, the design of the new TDMA system benefitted from the project experience. This work, described in the remainder of this section, has continued in collaboration consisting of RAL, Loughborough University of Technology, Manchester University, and GEC Marconi Ltd in the UK and as part of a broader collaboration in Europe called SATINE with participants in both Austria and Italy. European tests using the European Communications Satellite, ECS II, are continuing at the time of writing.

4.5.1 *Lessons from the project*

The weaknesses in the satellite access equipment used in the project were well known at the start of the project but it was clear that there was insufficient time to develop a new design and use this in the network before OTS ceased operation. Nevertheless a subproject was set up within the main project to produce a specification and design for a new, high-performance satellite access controller. It was hoped to construct and test a prototype of this controller during the later stages of the project.

A specification for a new satellite bridge, including access controller, encoding scheme and modem, but not including a new earth station design, was developed by a design group from the project. The specification, which was influenced by work carried out for the DARPA Wideband Satellite project by Linkabit Corporation, USA, included

- an improved TDMA controller, allowing modem and encoding rates to be varied dynamically within bursts. The controller would also handle block fragmentation across bursts and be capable of handling real-time traffic, such as voice.
- a high performance digital modem capable of variable bit rates within bursts at 1, 2, 4, and 8 Mbit/s.
- a codec using block coding suitable for implementation in VLSI, giving encoding rates of 1/2, 3/4, 7/8, and 1.

It was envisaged that the controller would interface to a variety of LANs or to digital circuits. The controller would also be given information about the quality of service required for different types of traffic and would use the variable encoding and bit rates to obtain this quality in the prevailing transmission conditions.

4.5.2 *The satellite access controller*

It was known that the 'serial' controller configuration shown in Fig. 4.2, with a computer-controlled interface in series with a link driving computer, was unsatisfactory. In this configuration both the computers have to deal with four data streams, two input and two output. In addition control information, which has to pass between the two computers, has to be mixed in with the data streams. In the new controller a 'parallel' configuration is used with one microprocessor system controlling the up-link and the other the down-link. The configuration is shown in Fig. 4.7. The parallel arrangement has the great advantage that, although each processor has to deal with four channels, only two of these are for data, the other two being for the low-volume activity of exchanging control information between the two processors. There is also the additional advantage that this control information is now separated from the data flow. Use of a minicomputer in a satellite access controller tends to force the use of the serial configuration to avoid the expense of using two such computers.

Detailed design and construction of a new controller was carried out by GEC and RAL. This controller uses two Multibus systems each with a Motorola 68000 processor and memory together with purpose-built DMA boards for access to the modem and earth station. A direct processor-to-processor link is provided for the limited communication necessary between the two halves of the controller. Use of the Multibus standard ensures that

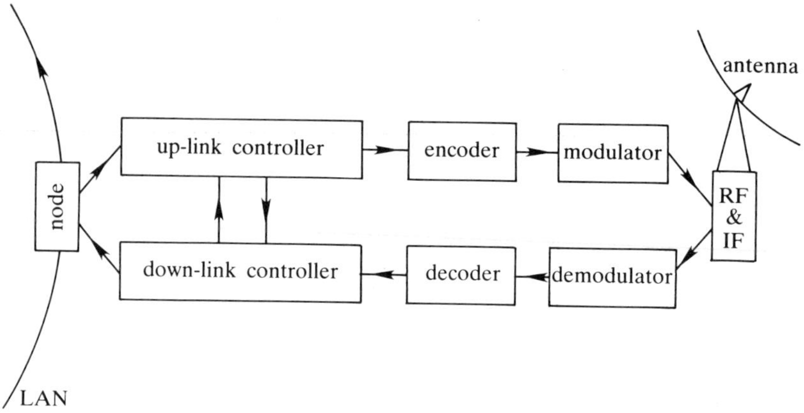

FIG. 4.7 New satellite access controller configuration.

interface cards are available to allow connection to a variety of LAN types and other communication media on the terrestrial side of the controller. The special-purpose DMA boards on the satellite side of the controller have been designed to enable the controller to handle small packets and packet fragmentation with high performance.

A prototype controller was built during the project and some preliminary tests with the earth station and 2 Mbit/s modem used in UNIVERSE were made in the last days of OTS at the end of 1983. These tests indicated that the controller would have a throughput performance of several hundred packets per second, rather than the 40 packet/s of the satellite bridge used in the project.

4.5.3 *The modem and encoding schemes*

The design and construction of a new modem for the satellite bridge was carried out at Loughborough University. The general characteristics of the modem are as follows

- the basic speed is 8 Mbit/s but this speed can be changed instantly, that is, between bits, in both transmission and reception to lower speeds of 4, 2, and 1 Mbit/s.

- the implementation is almost entirely digital, leading to a robust and predictable performance that is not subject to drift with time. It also allows the modem to be programmed to operate using different modulation schemes.

- the modem has fast methods for synchronizing the demodulator at the start of a signal burst. This allows the guard times between timing windows in the TDMA scheme to be kept short.

A requirement of the specification for the new controller was that it should be possible to mix information coded at different rates within the same transmission window. Thus, for example, within a transmission window it would be possible, say, to mix uncoded voice with half-rate encoded control information and information from a file transfer. For this purpose block codes seem more convenient than convolutional codes. Manchester University undertook to investigate suitable block codes and to develop algorithms amenable to implementation in VLSI.

By varying the transmission and encoding rates the access controller should be able to obtain optimal use of the satellite for different types of data. It is anticipated that this facility would be made available by quality-of-service parameters at the network access level of an overall communications architecture. Another potential application of the variable rate capability of the controller would be in helping to maintain quality of service in adverse weather conditions: bit-error rate performance could be maintained at the expense of reduced total throughput. In achieving this goal, the ability to vary rates in the controller would be a valuable addition to other mechanisms, such as the ability to increase the transmitted power, available to an earth station control system for maintaining performance.

5. The UNIVERSE network

5.1 The network

The preceding chapters have described the principles of the overall architecture of the UNIVERSE network, the technologies of the constituent segments out of which it was constructed, and the implementation of the satellite bridge. This chapter describes the network built during the project and used to carry out the set of experiments described in subsequent chapters. In this first section, its general configuration and construction are dealt with. A number of network components (such as bridges, name-servers, etc.) were specially constructed as part of the project and merit individual description: these are covered in the next section. Only a general view of the quantity and variety of hosts connected to the network at each site is given here; details of relevant hosts and, where necessary, their attachment to the network are described briefly in the next chapter, which forms an introduction to the experimental programme and gives details of the additional, common infrastructure needed to carry it out.

Considerable use was made throughout the project of the UK Joint Academic Network (JANET). Some reference to this has already been made in Chapters 3 and 4. Its role was primarily one of support which was invaluable. This chapter includes a summary of the various points of connection with JANET and the different ways in which the support was provided. As a point of history, JANET was officially born during the project: earlier connections were to the SERC network, which was subsumed as a constituent of JANET. No distinction is made here and the term JANET is used throughout.

5.1.1 *Configuration*

The UNIVERSE network interconnected a total of seven sites in England: British Telecom Research Laboratories (near Ipswich); Cambridge University; GEC's Marconi Research Centre (near Chelmsford); Logica's offices in London; Loughborough University of Technology; Rutherford Appleton Laboratory (in Oxfordshire); and University College London. Six of the sites were equipped with satellite bridges for direct interconnection via the satellite network segment. The remaining site, Logica, was less than 2 km from UCL and was connected to the network by a special, dedicated terrestrial link provided by BT. For the purpose of public demonstration at

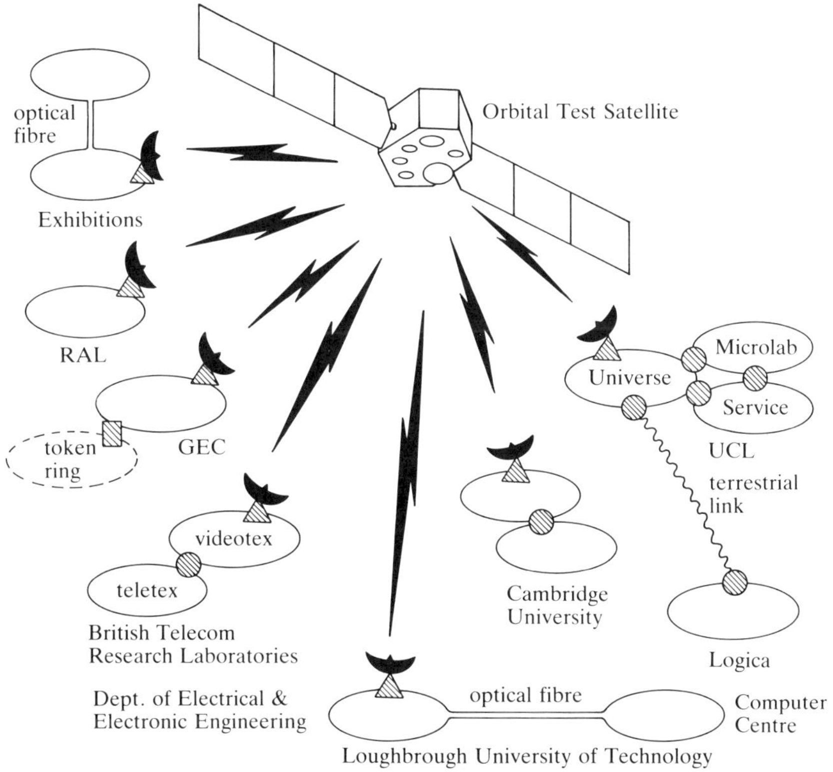

FIG. 5.1 The UNIVERSE network.

three exhibitions (two in London and one in Geneva), there was sometimes an eighth site, which used a mobile earth station, also provided by BT. The general arrangement of the network is shown in schematic form in Fig. 5.1.

As indicated in the diagram, the local network configurations at each site were generally different and the majority of sites had more than one LAN. The network element linking two adjacent rings within a single site was known as a ring–ring bridge: a short description of this is given in Section 5.2. Conceptually, the Logica–UCL link may be viewed as a ring–ring bridge in which the two halves have been physically separated and made to communicate via a cable instead of shared memory and CPU. The implementation of this arrangement is also described in Section 5.2. The inclusion in the project of two sites sharing use of a single earth station was considered important in the project, on both financial and practical grounds. The installed cost of an earth station with 3 m antenna is high; moreover, in urban environments, avoiding interference with other communications services in the area may make siting an antenna close to its users difficult or even impossible. The situation of Logica and UCL was typical in this respect and could be expected to occur quite often in service installations based on this technology.

The nature of the configurations found at each site was determined by the activities at each institution and the physical location of the participating groups within an institution. Only two of the participants had rings installed before the project began. One of these, of course, was Cambridge, where the number of devices connected to the ring had reached about 80 and the need to split the ring had become urgent. After the split, both rings still carried, in general terms, the same sort of traffic, namely that needed to support the overall operation of the distributed computing system developed there (described in Chapter 6). At UCL, the other site using the Cambridge ring before the advent of the project, the distribution of activity over the three rings (see Fig. 5.1) was along project or functional lines, as the names of the rings imply. There was a service ring, concerned generally with providing access to and communication among a number of machines used in a variety of service roles. The 'microlab' ring supported microcomputer-based teaching and development activities. The third ring was specifically for use in UNIVERSE. This separation of roles afforded each of the three activities a measure of protection from each other. It will be noted that the three ring–ring bridges at UCL formed a loop in the network topology. This introduced some additional complexity into the topology of the network and served to check that the architecture could cope, at least minimally, with such a configuration, though no developments were ever made to exploit it. It also offered a potentially more realistic basis for monitoring work than the simple (rootless) tree topology which otherwise would have existed.

Apart from RAL, which had just begun to use a Cambridge ring, the remaining sites did not have rings and these were installed as part of the project. At BTRL, two separate groups participated in the project. One of these groups was concerned with teletex service development, the other with videotex. It was deemed convenient for each to have its own ring and to link these by a ring–ring bridge. Two departments participated at Loughborough: the Department of Electrical and Electronic Engineering and the Computer Centre. They were situated in different parts of the campus and were about as far apart as Logica and UCL. In this case, a single ring was installed, with optical fibre segments between the departments (see Section 5.1.3). Such links were also used in all the demonstrations, where the earth station always had to be sited separately from the rest of the equipment. The most satisfactory way of linking turned out to be to site the satellite bridge close to the earth station and to extend the ring by means of a pair of optical fibre segments, as indicated in Fig. 5.1. The final variation in site configuration was at GEC, where a token ring was developed (see Chapter 9) and connected to the Cambridge ring there, albeit by a higher-level relay.

Reference has already been made to the three public demonstrations of the network, which formed a subsidiary project with its own partial funding. This activity is described in Chapter 10. Of relevance here is that it required the provision of a further (mobile) site, complete with its own Cambridge ring, satellite bridge, and earth station. This brought the total count of satellite bridges to seven. The network interconnected 12 Cambridge rings and one token ring and incorporated five ring–ring bridges.

During the last six months of the project, in the latter half of 1983, the number of hosts interconnected by the network reached some 200. Of these, about 100 were connected as a direct consequence of the project; the remainder were mostly situated at Cambridge and UCL and served purposes other than project UNIVERSE. Quite a wide range of machines was connected. The larger systems included an IBM 3081 at RAL; PDP 11s at Cambridge, UCL, and Logica; GEC 4000s at GEC, BTRL, and RAL; Primes at Loughborough and Logica; and a VAX and the CAP machine (Wilkes and Needham, 1979) at Cambridge. Microcomputer systems were connected everywhere on the network and were the main hosts used in the project. They included systems based on 6502, Z80, 8086, and 68000 microcomputers. More details about the hosts used in the project are given in Chapter 6, which describes the common infrastructure used for the experimental programme, and in later chapters where particular hosts played specific roles in some of the experiments.

In order that the network could function according to the architecture described in Chapter 3, every LAN required a nameserver. The only exception was the GEC token ring which was linked at a different level. Thus 12 nameservers were required. Each site except Logica was also equipped with a bootserver, so that small machines could be loaded in a uniform fashion via the ring: this required a total of seven bootservers. Both of these systems are described briefly in Section 5.2.

The UNIVERSE network was highly experimental: even when all sites were scheduled to be operational and all equipment was working everywhere, a state which was only even approached during the third and final year of the project, the satellite segment was scheduled to be available for only three hours each day. It is doubtful if the UNIVERSE network could have been built at all in the time available without the aid of an existing network to assist the development and maintainance of the satellite segment and to provide a permanently available communication infrastructure to aid collaboration.

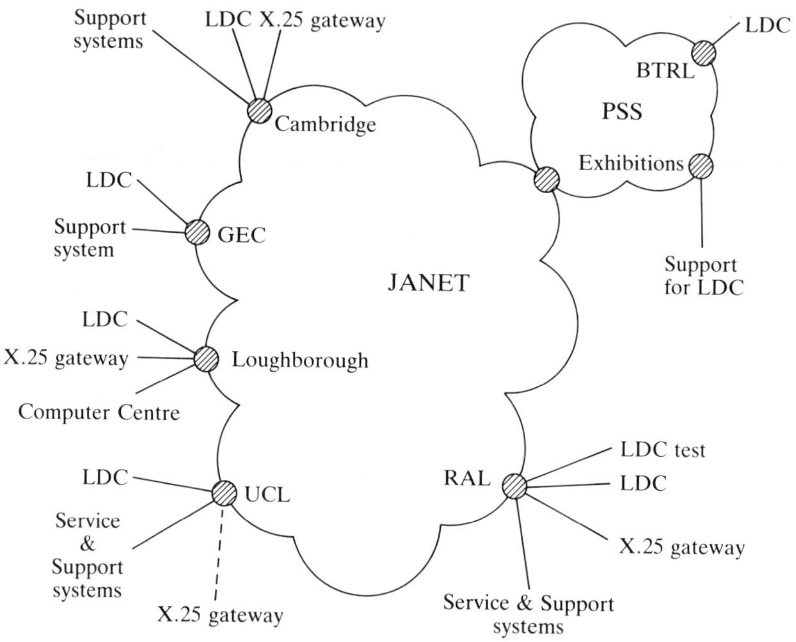

FIG. 5.2 Use of JANET by the project (schematic).

5.1.2 *The role of JANET*

The network which provided this communication infrastructure service to the project was JANET. For communication with BTRL and the demonstration site (Chapter 10), this was augmented by use of British Telecom's Packet Switched Service (PSS). These two networks were interconnected by a 'gateway' which enabled JANET-style transport connections to be made end-to-end, as well as providing conversion between TS29 and X.29 (CCITT, 1981*b*) and X.29 relaying. For the purposes of this project, it was possible to regard PSS as a transparent extension of JANET.

In terms of usage, three sorts of connection could be distinguished at the project sites, though not every site exhibited all three. Firstly, each LDC was connected as a JANET host. The purpose of this is described in Chapter 4 and it suffices to reiterate here that this was an invaluable tool in speeding up all stages of development, testing, and fault diagnosis of the software for the satellite bridge. Secondly, during the project, a so-called X.25 gateway which provided JANET-style 'Yellow Book Transport Service' (PSS User Forum, 1980) relaying was developed. The function of this gateway from a network architecture standpoint is described in Chapter 3. In principle, it extended the functionality of JANET into the UNIVERSE network; moreover, when the satellite segment was not available, those sites with X.25 gateways could maintain this level of connectivity with each other. The so-called 'basic block tunnel' mentioned in Chapter 4 also relied upon the X.25 gateway to obtain the transport connections it needed to connect to other sites. The X.25 gateway and the basic block tunnel are both described in Section 5.2. Thirdly, sites used JANET to interconnect their computing infrastructure support services to each other. The general arrangement of the JANET connections of relevance to UNIVERSE and their usage by each site is shown in schematic form in Fig. 5.2.

The existence and form of what has been termed here the 'computing infrastructure support services' varied from site to site. The types of service under consideration here include those for developing and debugging software to run on a variety of target systems; preparing documents and making them available to others for reading or collaborative authorship; use of electronic mail locally on one site, among the project participants, and more generally in the community; means to access local or remote machines in the community. All of these facilities were used in the project. Mail was especially useful as a regular means of assisting collaboration. Specific uses included technical discussion, arranging meetings, publishing details of the satellite schedule, etc. Together with file transfer, it was also valuable for transferring software between sites. Terminal access to remote systems was invaluable for a variety of purposes, including remote testing, and inspection of document and software source files. The usefulness of this infrastructure was much increased if a site had any form of internal connection between its infrastructure services and its segment of the UNIVERSE network, as was the case in some fashion at Cambridge, GEC, RAL, and UCL, which all had such support services in one form or another. Such connections greatly assisted the sharing of software for network and applications systems. Their absence in varying degrees from the other sites was on occasion a disadvantage which was felt particularly at Loughborough in respect of software sharing.

5.1.3 *Construction*

Construction activities in the project were organized along a number of different lines. Responsibility for areas of work was of course delegated by the committee of principal investigators to one of the working parties listed in Appendix A. The construction and operation working party was given the overall charge of building the network. The largest single activity here was the construction and installation of the satellite bridge and was correspondingly the main preoccupation of this working party. Responsibility for the provision of the individual site network segments remained with each site. Production of individual components for use throughout the project was usually undertaken by a particular institution acting under the direction of one of the working parties. Thus, RAL provided the satellite bridges, Cambridge provided the nameservers and ring–ring bridges, and UCL provided the bootservers. Organization of the provision of the link between Logica and UCL was the province of the so-called cables working party; the cable itself was provided by BT and the half-bridges at each end were provided by Logica. The provision of nameservers and bootservers was part of a more extensive exercise in replicating selected parts of the Cambridge system and was the reponsibility of the distributed computing working party.

Although some sites already had Cambridge rings installed and in use, the majority had not and so construction of the network included ring installation. Almost all the ring equipment (monitor stations, repeaters, stations, and slave power supplies) was obtained from either Logica VTS Ltd (the Polynet range) or Scientific and Electronic Enterprises Ltd (SEEL, Transring 2000 series). At Loughborough, the participation in the project by two departments, separated by a kilometre, gave rise to the need for long ring segments. This led to the development by BICC, Loughborough, and Logica VTS of a product (called the Polynet Optlyne unit) marketed jointly by BICC and Logica VTS. The unit used Manchester encoding to achieve coupling of two ring nodes by a single optical fibre. This product, used initially at Loughborough, was used at all the exhibitions (Chapter 10), where its immunity to interference proved important. It was also used later at RAL to extend the ring across the site.

The importance to the project of common or intersecting membership of working parties has already been remarked upon in Chapter 1. Nowhere was this more important than in building the network; three working parties have already been mentioned—construction and operation, cables, and distributed computing—all of which had more or less direct responsibility for aspects of network construction. In fact, two more of the working parties were also of importance in this phase of the project. All questions relating to the network architecture and protocols, their implementation and documentation were dealt with by the protocols and implementation working party. Use of JANET by the project was organized by the X.25 working party; this included the connection of the LDCs to JANET, which was so important for satellite bridge development and testing.

The overall timescales of the project have been described in Chapter 1. The planning and critical development necessary for construction of the network were largely completed in 1981; the bulk of the actual construction took place in 1982. Of the various components of the network, detailed description of the satellite bridge has been given in Chapter 4; the remaining components are described in the next section. In common with most projects of any size, a wide range of components from a variety of manufacturers was used by each of the participants in constructing those parts of the system for which they were responsible. An example was the use of the ring equipment from several vendors at a time when this was not normally done and there were no standards. Another example was the use of several different bus systems for equipment construction. While there were good arguments in each case, the net result was that the project became critically dependent upon having a very wide range of expertise either within, or very easily accessible to, the project team.

Indeed, although the range of electronic, computing, and communications activities indulged in by the project was quite broad, the use of 3 m antennas even caused a number of excursions into the field of civil engineering. These antennas were variously sited on roofs, at the top of a lift shaft, and in the centre of London. Issues such as safety (from radiation during transmission), lightning protection, structural stability in high winds, and siting to avoid radio interference all required attention and took effort to resolve.

5.2 Components

In describing the realization of the UNIVERSE network architecture, reference has been made to a number of individual components whose presence was essential to the functioning of the network, including its interconnection with JANET. This section completes the exposition of the UNIVERSE network, its architecture and realization, which has been the theme of the first half of this book, by describing the general construction and operation of those components.

The components in question are the ring–ring bridge; the Logica–UCL link (referred to as the cable link bridge); the nameserver; the bootserver; the X.25 gateway; and the basic block tunnel. The first three of these were primary components of the main network: they formed part of the implementation of the architecture of Chapter 3 and without them the network would not have existed. The bootserver provided support at several levels. Although labelled 'bootserver', it provided name table loading services for the nameserver and site logging facilities, as well as a general booting service, which was in particular used by the ring–ring bridges. The X.25 gateways were a means by which the network was connected to JANET at several sites. The basic block tunnel made use of the network relay services provided by the gateways to emulate the site interconnection service provided by the satellite segment and was intended for use when the latter was not present.

Three of the components just enumerated are forms of bridge. They all differ in various ways, both from each other and from the satellite bridge described in Chapter 4. They each provided the essential bridging functions necessary to the implementation of the UNIVERSE architecture. While those functions are implicit in the description of the architecture given in Chapter 3, it seems useful to begin the description of the components with a brief summary of these functions, in order to place in context the way in which they were provided by each form of bridge.

5.2.1 *Bridge functions*

Bridges have to handle three sorts of client traffic: datagrams, SSPs, and LVCs. Although the processing for each of these is different, a common scheme is used to establish the route to be used to reach a global address. Whereas nameservers alone are responsible for the binding of names to addresses, bridges and nameservers share the task of implementing the fixed routing used in UNIVERSE. The global addressing scheme is explained in Chapter 3 and consists of four components: *site*, *subnet*, *host*, and *port* (see Fig. 3.3). At each site a fixed set of routes linking every subnet to every other on that site is defined; routes are also defined between each subnet and the single bridge to be used for all off-site traffic. Given this information, a bridge can determine from a global address which (if any) is the next bridge in the route to that address. All bridges obtained this site routing information when they started up. No provision was made for dynamically altering it thereafter. It should also be noted that Logica was, for addressing purposes, the same site as UCL: the ring at Logica was simply a different subnet.

In all cases, the first (perhaps only) 'hop' in a route is determined by the nameserver: only in the case of traffic to another subnet (or site) is a bridge involved. In the latter case, for SSPs and LVCs, the nameserver makes an address insertion request to the first bridge in the route and supplies the global address of the destination to the bridge. The bridge allocates a port and returns this to the nameserver, which returns it and the station address of the bridge to the client. For datagrams, the port on the bridge is fixed and no such interaction takes place. Datagrams contain a global address from which each bridge determines either that it can send it to its final destination (same site and subnet) or that it is on a different subnet (or site) and has to be forwarded to another bridge.

For SSPREQs and OPENs, the bridge converts these to BRIDGESSPs and BRIDGEOPENs, respectively, by adding the global address and, using the same routing algorithm as that just described for datagrams, forwards them to the next bridge, having inserted its own reply port. This process of forwarding and establishing the reverse path continues until the last bridge is reached, which converts back to SSPREQ or OPEN and sends this on to the destination. All bridges in the path set appropriate time-outs and await replies. An SSP reply (SSPRPLY) may be followed by data blocks along the reverse path, but no forward path is established, whereas for an LVC paths are established in both directions. A reply with an unsuccessful return code causes the reverse path to be deleted and, in the case of an LVC, no forward path is created.

The arrangement is asymmetric in that OPENACK and SSPRPLY transmissions are 'in-band' (i.e. along the path to be taken by data) and involve only the bridge. The OPEN and SSP request both achieve separation from the data paths by 'out-of-band' signalling, which is shared by the nameserver and bridge.

Neither an OPEN nor an SSP request contains the address of the originating client. In a single ring environment, the availability of a service provided by the nameserver to map a station address to the name of that station enabled a server host to discover who its client was. This is not sufficient in the UNIVERSE architecture, since the best which could be provided by such a service in the case of a remote client is the name of the proximate bridge (to the server) in the path. A new service was defined, called *reverse trace*, in which a request was made of the nameserver, giving both a station and a port. In the case of the local nameserver finding that the local station is a bridge, it sends a BRIDGETRACE request to the bridge, which is forwarded back along the bridges in the path to the first bridge, which in turn makes a *reverse global lookup* request to its local nameserver and forwards the result back to the original nameserver, which can now inform the host making the original request.

5.2.2 *Ring–ring bridge*

Investigation into possible means to interconnect Cambridge rings within a single site began at Cambridge University before the UNIVERSE project was proposed. The result was a ring–ring bridge (Leslie, 1983) which was used at three of the sites participating in UNIVERSE: Cambridge, UCL, and BTRL (see Fig. 5.1). Like the satellite bridge, much of the critical development work was accomplished with funds committed from sources outside the project, so that the development reached the prototype stage during 1981, ahead of the completion of planning and the beginning of construction for UNIVERSE.

The bridge provided all of the functionality described in the previous section. The whole system was entirely symmetric, reflecting the symmetry in the bridge functions in each direction between the two rings. Consequently, the implementation was in terms of two *bridge halves*, one for each direction through the bridge. These halves called upon common services for such activities as port management, timing, and monitoring. (*Note* The use of bridge half here should not be confused with the term half-bridge used elsewhere to refer to the systems at each end of the Logica–UCL link.)

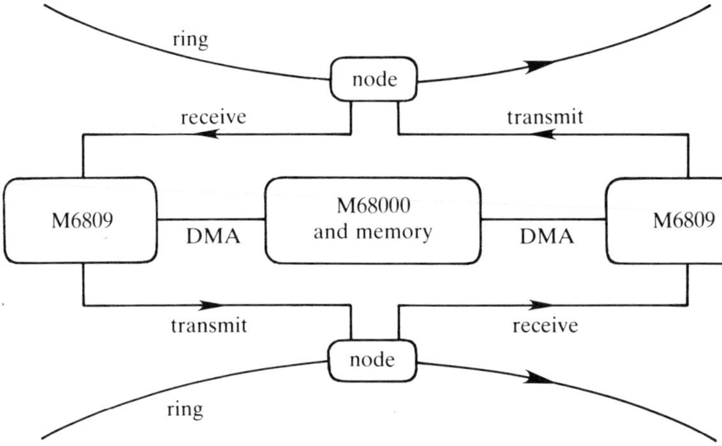

FIG. 5.3 Ring–ring bridge component configuration.

The ring–ring bridge was constructed around a Motorola M68000 processor with 256 Kbytes of memory. Attached to this were two intelligent ring interfaces, each containing a Motorola M6809 processor and DMA channel to main (68000) memory. Each of these interfaces received on one ring and transmitted on the other, so that they effectively cross-coupled the rings being connected. Main memory was used for all reception and transmission buffers, via DMA. This arrangement, shown in Fig. 5.3, meant that the majority of traffic only passed through a single interface and could transfer from one ring to the other without passing through the central 68000. This allowed buffer management to be performed by the interfaces rather than by the 68000. The individual physical components used in construction of the ring–ring bridge were the same as those in the large server described in Chapter 6. The crate was also similar but with a different backplane arrangement.

The system running in the central processor was based upon the TRIPOS operating system (Chapter 6). The code in the two 6809 ring interface controllers was identical but peculiar to the bridge. The main tasks of interest were the two bridge halves, a low-priority port timer task, and a high-priority port manager. There was also a TRIPOS device driver for each ring interface controller, which was managed by its own bridge half. The bridge halves contained the majority of the code in the system, and were responsible for routing and setting up port mappings. The port timer task decremented the timers on each port mapping, while the port manager was responsible for allocating and deleting ports. Since the bridge halves kept a small pool of free ports, the most common interaction for a block of data involved only two TRIPOS task switches: from an interface device driver to a bridge half and from the bridge half back to the (same) interface device driver. This was true also for OPEN, OPENACK, and SSP request and reply blocks.

The ring–ring bridge system had no integral disk storage and on start-up the two ring controllers, executing identical code in programmable read-only memory (PROM), co-operated to attempt to boot the central 68000 processor from bootservices on the rings on either side of the bridge. One side having succeeded, the controllers then waited, having started the central processor. The latter then loaded both controllers with bridge service code incorporated as part of its own boot image. The final stage of starting up consisted of enquiring of the *read routes* services (Section 5.2.4) on the nameservers on both sides to find out the identity of the subnets on each side and the (fixed) 'next hop' information for routing to any subnet or site in the network.

Besides the main tasks of the bridge, there were a number of auxiliary ones. There was a console handler (and a pair of console device drivers), a resident debugging task, some performance monitoring tasks, and a root task which provided a console interface to handle queries about the state of the bridge and could also be used to display traffic statistics. The performance monitoring included both network monitoring information and internal observation of CPU usage.

For full details of the performance of the ring–ring bridge Leslie (1983) should be consulted. Using 2 Kbyte basic blocks, the bridge could forward data from one ring to the other at a rate of about 220 kbit/s. If data was being forwarded in both directions simultaneously, the rate in each direction dropped to about 200 kbit/s. These figures demonstrate that there was relatively little interference between traffic in opposite directions and illustrate the success of the basic design configuration. Investigations into how to increase throughput further would need to include precise determinations of current resource utilization, including all three CPUs and the backplane bus. This was not done during the project.

5.2.3 *Logica-UCL link*

The overall objective of the project undertaken by Logica to link UCL and Logica by terrestrial cable was to provide a connection between the rings at each end which was functionally the same as a single bridge (Section 5.2.1). From the standpoint of the rings at each end of a chain of rings connected to each other by ring–ring bridges, there is no functional difference between such a chain and a single bridge. The possibility of using elements of the cable link to provide a ring and to achieve the overall connection by

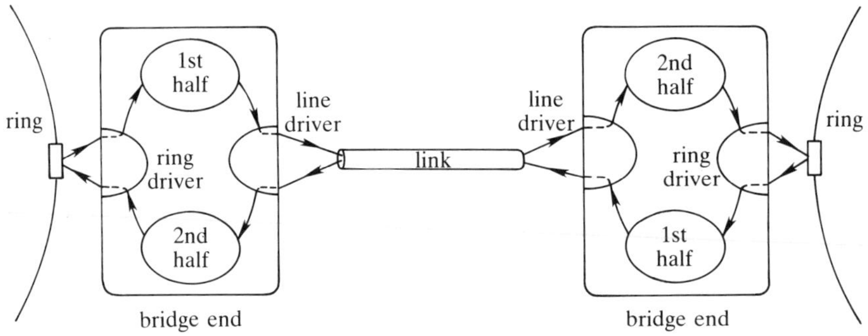

FIG. 5.4 Principal processes in bridge end systems (schematic).

placing a ring–ring bridge at each end was discussed in the project. This was not pursued partly as a result of the difficulty encountered in finding out the precise properties of the cable link with which the project had been presented and partly because of regulatory problems associated with using such a link to carry so unusual a signal as that needed for a Cambridge ring. In the eventual design, as might be expected of systems not sharing memory, the systems at each end of the link (called ***bridge end systems***) did not share state, and their functionality was closer to that of a pair of ring–ring bridges, suitably optimized, than to a single bridge conceptually separated into two halves. (*Note* The bridge end systems referred to here are the same as the half-bridges already mentioned informally on pp. 28 and 86.)

In describing the operation of the bridge end systems, the terms ***first*** and ***second bridge halves*** are used (not to be confused with half-bridges (see note above) nor with the bridge halves used in describing the ring–ring bridge in Section 5.2.2). Each bridge end system contains one of each kind of bridge half and traffic in a given direction is processed by the first bridge half in the first end system and by the second bridge half in the second end system (see Fig. 5.4). The optimizations referred to above in comparison with two full bridges arise from self-evident properties of the link: the first half at one end need not perform any routing to send packets to the second half at the other end; similarly, the second half does not have any need for its own set of ports when receiving from the first half.

Briefly, the bridge end systems operated as follows. As explained in Chapter 3, a feature of the network architecture is that during the initial exchange of messages, a control message travelling in one direction sets up the path in the opposite direction. This applies to SSP, OPEN, and OPENACK (and the corresponding forms used between bridges). The implementation of the cable link bridge is instructive in that the division of tasks between

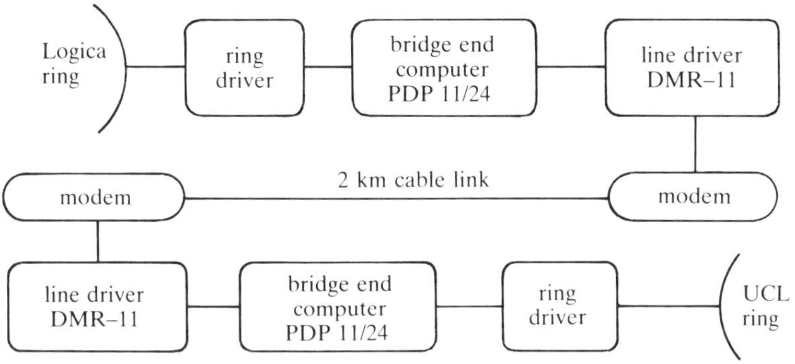

FIG. 5.5 Cable link bridge components.

the two ends and the two halves at each end precisely reflected this aspect. In gross terms, the first bridge half handled address insertion requests, setting up port mappings (by holding the tables, assigning the port for use in receptions by the first half at the other end, and inserting this port in the reply or data port fields of all forms of SSP, OPEN, and their replies), and conversion of SSP and OPEN to BRIDGESSP and BRIDGEOPEN. The second bridge half performed routing, either by interpreting a global address (datagram, BRIDGESSP, or BRIDGEOPEN) or by using the mapping established by the first half in its own end (SSPRPLY, OPENACK, or data). In order to achieve the latter, the (reply or data) port on which the first half received the block was appended to the block for transmission to the second half in the other bridge end: this port was then used as a handle by which to discover the port and station to which the block should be sent. A point to note is that the reception port, assigned by one end and used by both ends, was the unique handle by which paths were identified.

The equipment used to implement the link between Logica and UCL is shown schematically in Fig. 5.5. The principal item at each end was the bridge end computer, which was a PDP 11/24 with integral disk file storage. It used a Logica proprietary interface to the ring and a DEC DMR-11 line driver; a proprietary modem provided the connection to the 2 km link. This arrangement could in principle sustain data rates of up to 1 Mbit/s. The bridge end system was implemented as a single task executing under the RSX operating system. It was identical at each end. In addition to the processes already described, it contained process modules for buffer management, error processing, and some performance monitoring.

The bridge end systems booted independently off their own disks. They did not use the routing information service provided by the nameserver but obtained this information from an internal file.

Errors could arise in a number of ways. Generally, they were caused by the bridge running out of resources, by hardware faults, or by the user sending erroneous blocks. The general response was to discard the block in error and to record this fact by sending an error report to the console of the bridge end computer and attempting to send an 'error report' block to the other end. These reports could also be sent to a file and to the local site log service (Section 5.2.5).

Only a few preliminary performance measurements were made by sending global datagrams from the PDP 11/44 on the Logica ring via the bridge to the LDC at UCL. The results suggest an upper limit of around 120 kbit/s for maximum length blocks, although in practice only 1 Kbyte blocks could be sent. The observed, practical limit was about 100 kbit/s; allowing for an estimated 20 per cent overhead, this amounted to only about 12 per cent utilization of the available bandwidth of the link.

5.2.4 *Nameserver*

Together with the concept of a bridge, the nameserver plays a key role in the UNIVERSE network. Its purpose in the network architecture has been explained in Chapter 3. In order to realize a practical implementation of the network architecture, additional functionality was required of the nameserver, over and above that needed to fulfil its role in the basic architecture. This has already been touched upon in Section 5.2.1 in describing the mechanism by which nameservers and bridges co-operate to provide network servers with the reverse trace facility for identifying their clients. The role of the nameserver was also extended in one or two other ways, for example, to act as a repository for the fixed network routing information needed by bridges and to provide assistance in inter-domain address mapping needed for interconnection with external networks. This section completes the picture of the nameserver given so far by describing its full functionality and the way it was realized in the project.

For the purpose of implementing the necessary services, the functions of the nameserver can be divided into a number of categories:

- name table storage;
- responding to queries about the contents of name tables;
- co-operating with bridges and other nameservers to create or trace paths according to the fixed routing scheme of the network; and
- table management.

Conceptually, the nameserver provides access to three separate, local tables: a *domain* table, a *routing* table, and a *name* table. For comparison with the original nameserver operating in a single ring environment, the first two of these are additions peculiar to the UNIVERSE architecture; they are also small in comparison with the name table. The domain table has an entry for each domain name, the contents of which are a site and subnet address. Since all nameservers are at a known, fixed station number (240), the domain table effectively defines both the extent of the network known to the nameserver and nominates a nameserver to be contacted for lookup in each 'foreign' domain. The entry in the domain table for the nameserver's 'home' domain, while logically present, has somewhat different significance from the other entries and is handled in a quite different way. The significance of this home entry is that it defines the location of the nameserver in the network.

The routing table effectively maps every subnet in the network to the address of a bridge, on the nameserver's own subnet, by which the remote subnet may be reached. This is logically a two-stage arrangement: a single bridge (which may not be on the same subnet as the nameserver) is nominated for all off-site communication and a bridge is nominated for traffic to each of the other subnets on the site.

The name table is implemented as a database with only one type of entry, called a *name entry*, and is assumed to be small enough to be held in main memory at all times. The fields of a name entry are name, site, subnet, station, port, function code, and transformed string. The primary purpose of the table is to map the names of services to service addresses. It also serves to map the names of machines (and bridges) to station addresses.

The name table is divided into two parts: local and foreign. Names without a domain component are local, by definition; all local names are stored in this form. Once loaded, the local part of the name table is fixed; the foreign part, however, is defined to be a cache. The reason for this division resolves to one of administrative control. It is assumed that the local part of the table will always reflect the true state of affairs in the home domain, since both are presumed to be under the control of the same administration. However, in general, no assumptions may be made by the administration of one domain about the mapping of names to addresses in another domain. Thus the entries in the foreign name subtable are liable to invalidation without notice and arrangements to refresh need to be made (see below). It should be noted, however, that it is perfectly possible to define a local name which refers to a service in a foreign domain. While this can be quite useful for experimental operation and was indeed quite widely used for this purpose in the project, it is not at all suitable for general operation because the service may be moved without notice by the foreign administration and there is no automated remedy.

The majority of entries in the name table refer to services. Those which are the names of machines or bridges are flagged and do not contain a port or function code. The function code was mentioned briefly in Section 3.2.3.

It has no role in the network architecture: it simply allows a system to offer a variety of services through the same well-known port. The transformed string field in an entry enables the the nameserver to provide a simple mapping service from names in the UNIVERSE network domain to other naming domains. The only regular use of this facility was in connection with the X.25 gateway to JANET and is described in Section 5.2.6.

All the services of the nameserver are accessed by SSP transactions. Three basic facilities, *list names*, *list domain names*, and *read routes*, enable the contents of the three tables to be listed. All three have obvious use in table management. The *read routes* service allows a bridge to obtain its routing information from the nameservers on each of the subnets to which it is connected. A *reverse lookup* service is provided to enable a client to find the local name of another system given its station address. Two services, *own name* and *own address*, enable a client to determine its own name and address. Both of these assist in the creation of portable code. The latter service may also be regarded as compensating for a deficiency in the ring station hardware which has no facility to allow its host to determine the address setting of its own ring node. Finally, two 'global' services are provided: *global lookup* and *reverse global lookup*. The *global lookup* service returns the complete, global address of a name, provided the name is present in the nameserver's tables (it does not consult other nameservers). Whilst this service is available for use by any client, its essential purpose is to provide other nameservers access to name table entries. The *reverse global lookup* provides the full UNIVERSE name (i.e. domain and local name components) of any station in the nameserver's tables.

The two services most important to the operation of the network are *name lookup* and *reverse trace*. The first of these is essential to the implementation of the network. Its role in relation to the network architecture has been described in Chapter 3. For names in the home domain, operation is straightforward: for a service on the same subnet, the result is essentially just the contents of the name entry (excluding site and subnet); otherwise, an address insertion is performed first on the relevant bridge. In the case of an unknown foreign name, the nameserver performs a global lookup on the nominated nameserver in the foreign domain and caches the result for future use. The rest of the operation then proceeds as before. If the foreign name is already in the cache, the nameserver proceeds immediately to any necessary address insertion and then responds to the client. The architecture specifies that it should then immediately perform a global lookup to revalidate the cache, so that subsequent retries by the client, caused by an invalid cache entry, will succeed.

The purposes and mechanisms of reverse trace have been explained in Section 5.2.1. While not essential to the basic implementation of the network architecture, certain higher-level communications and applications do require it. Examples are communication authenticated by address (see Chapter 7) and booting, where the boot service determines the image to load on the basis of who is requesting to be loaded.

The UNIVERSE nameserver was implemented by Cambridge on a Z80 'small server' system (see Chapter 6), configured with 8 Kbyte of ROM and 56 Kbyte of RAM. The code of the nameserver resided in ROM since without a nameserver the boot service cannot be located. Space is also reserved in the ROM for an initial name table. All unusual events, including name table reload, startup, or path setup errors, are recorded by sending a message to the site log service.

Management of the tables in the nameserver encompasses a variety of mechanisms. The most dynamic of these is the cacheing of foreign names (see pp. 95 and 96). The cache handling implementation departed from the specification. Instead of initiating a new lookup of a foreign name after its use from the cache, in order to refresh or correct the cached information, a cached name was given a fixed maximum lifetime of about two minutes after which it was discarded. This was adequate for use in the project but would obviously require correction in any less experimental situation.

The contents of the name table are read initially from the small table included in ROM but thereafter come from an external service called *name table read*. The initial table must contain the home entry in the domain list, which defines the domain, site, and subnet where the nameserver is located. Other entries in this table normally include those for name table read, the services provided by the nameserver, the site log, and perhaps the boot service.

Reloading of the name table from the external name table read service is initiated either by a manual reset of the Z80 or by sending an SSP request to the *nameserver reset* service. In either case, the new table is only loaded if its version number is more recent (larger) than the current one. This 'master' name table from the external service may contain routing information for a number of subnets: the nameserver stores only that information relevant to its own subnet. The purpose of this is to allow a single master table to be maintained for a given domain, as an aid to keeping all the nameservers within a domain consistent. Details of the provision of the name table read service and maintenance of the master table are included in the next section.

5.2.5 *Bootserver*

Many of the smaller systems used in the project had no backing storage and were arranged to boot from a ring service. It was the function of the bootserver to provide this service. The machines included ring–ring bridges, large and small servers (see Chapter 6), and several other systems, including X.25 gateways. Some other services necessary to support the network were also incorporated into the system which provided the booting service. These were the name table read service referred to in the previous section, a site log service, and a file access and transfer service.

Two protocols were used for booting. One of these predated the start of the project but could not be used to boot systems with more memory than 64 Kbyte. Reliance upon this protocol by a large number of systems effectively prevented any change to it. The second protocol was developed simply in order to cope with the new, larger systems which came into use during the project. In the interest of simplicity, neither protocol uses any flow control, on the assumption that the recipient can sustain rapid flow because the data is for memory and the network elements have sufficient buffering to support bursts of this order of magnitude. The newer protocol allows acknowledgements, primarily to keep the reverse path of the LVC from timing out so that positive confirmation of success could be sent at the end. The older protocol did not do this.

The site log was a straightforward service, based on SSP transactions from the network, displaying and filing text messages. It also provided a time of day service, again via an SSP interface. The files of error and status reports so generated were closed each day at midnight to assist retrieval and management of the information.

To reload its name table, the nameserver used a subset of the Cambridge fileserver protocol; at Cambridge, it read the table directly from the fileserver. The implementation of the name table read service at other sites thus consisted of providing a subset of the fileserver primitives. A means of accessing the file store on the system was also needed so that new boot files could be added and site log files accessed remotely. Meeting these requirements was combined by providing a simple, general purpose, fileserver interface composed essentially of read, write, lock, and unlock primitives.

The basic requirements of the system needed to provide the support functions described above can be summarized as follows. The facilities of the operating system and the size of main memory needed to be sufficient to enable a number of independent processes to execute simultaneously for long periods to provide the requisite ring services. A suitable means of interfacing the system to the ring was obviously required: the preferred solution here was for a system based on the Multibus, to enable use of the Polynet VMI-1 interface. The file store needed to be large enough to contain operating system and application service programs, boot images ranging from a few kilobytes to several hundred kilobytes, name tables, site log files, and a few application data files accessible via the fileserver primitives. A 10-Mbyte Winchester disk was considered the minimum requirement here.

A final necessity was the inclusion of a floppy disk drive. In the event of code or name tables being so damaged at a site that essential network elements could no longer function, relevant files could be transferred to the site on floppy disk to repair the situation.

The system chosen was a WICAT workstation, based on Multibus and a 68000 processor. It had 10-Mbyte Winchester and floppy disk drives. Implementation of all the functions described, as well as interfacing the system to the ring, were all carried out by UCL. While the system fulfilled its defined role in the project, there were some fundamental weaknesses which precluded its further use in the project. Chief among these were fragmentation and high occupancy of main memory by the proprietary operating system, which effectively limited the practical number of simultaneous user processes to about three.

Two lessons were learnt from this system. Firstly, it was a mistake not to have full remote terminal (login) access to such an important infrastructure element: the contrast with both the LDC (Chapter 4) and the 'large server' (Chapter 6) only served to emphasize this. Secondly, the importance to an experimental project of having adequate access to complete and expert information on the hardware and software of any key component cannot be over emphasized. Key expertise was available to the participants for most of the rest of the equipment used in the project: its absence in this case demonstrated the point clearly to the participants.

5.2.6 *X.25 Gateway*

The purpose of the X.25 gateway was to provide a general interconnection between the UNIVERSE network and the UK academic network, JANET. The architectural principles of this gateway or, more precisely, UK 'Yellow Book' transport service relay, have been explained in Section 3.6. There were originally several motivating factors which led to the development of this relay within the project.

- The provision of alternative inter-site paths of connection for control purposes which avoided use of the satellite segment under investigation.
- The ability to transfer software files between experimental systems on different sites, independently of the availability of the satellite segment, particularly for the preparation of experiments using the satellite.
- To enable terminal login connections to be made between systems on the UNIVERSE network and those on JANET, as well as between UNIVERSE systems on different sites in the absence of the satellite segment.
- As a prerequisite for the basic block tunnel described in the next section.

At the time of the project, although the technology to build such a gateway was well understood and the components existed, there was no suitable product available for purchase by the project.

In developing the design for the gateway, some consideration was given to the arrangements for (UK) transport service address transformation between the UNIVERSE network and JANET. The general principles of such transformations are set out in the UK 'Yellow Book' (PSS User Forum, 1980). Addresses are considered to consist of a sequence of components, each relating to a domain in a concatenated sequence of domains which is logically traversed to reach the target service, or *called address*. Typically, gateway implementations simply treat this logical addressing structure as a transport level source route address. Under these circumstances, the *calling address* begins as a null address and the first component of the called address is the *active* component, used to address the first gateway along the route to the target. As the process of connection proceeds, the calling address is built up and the called address components are successively activated and removed.

In this context, the UNIVERSE network formed a single domain. During the establishment of a call coming into the gateway from JANET, the active component was treated as a UNIVERSE service name and looked up in the nameserver. A generalization of the standard processing of addresses was introduced at this point: the transformed string returned by the nameserver was included as part of the next (usually last) active component of the called address when using the other results of the name lookup to set up a CR82 connection to the relevant UNIVERSE service. Though in principle a potentially interesting facility, for example, for relayed redirection of calls, no use was made of this in the project.

The same sort of arrangement can be used for calls in the opposite direction, that is to say, a component in the called address can be used to address the relevant JANET service. Without interpretation in the gateway, and in the absence of the equivalent of a 'JANET nameserver', this has to be a JANET X.25 DTE address. In order to address a JANET service by its transport service 'title' or name, a mapping service is required. At the time of writing, this information is kept in a database called the network registration service (NRS), accessible via the network, and individual hosts keep regularly updated caches of those parts of the database which they need. At the time of the project, no such service existed.

The mechanism adopted had the merit of simplicity and required little effort to implement. The gateway did not perform any mapping function: it was presented with a called address in the form of a string value whose first component was the JANET X.25 DTE address. However, applications using the service could make use of the nameserver to present a more acceptable interface to users of the application. This was done by making an entry in the name table whose name was the JANET transport service title, whose transformed string was the JANET transport service address, and whose UNIVERSE service address was that of the gateway service. A general name for this latter service was also available which had a null transformed string: by using this and supplying an explicit JANET service address, application users could address arbitrary services.

Finally, the entry for the gateway machine name, used to identify the gateway station address, also contained a transformed string which represented the JANET X.25 DTE address of the gateway. The latter needed this for insertion into the calling address parameter of connections made from the UNIVERSE network to JANET.

The X.25 gateway system was implemented by Cambridge and RAL. It consisted of a number of tasks which executed under the TRIPOS operating system on an 8086 Multibus system with 192 Kbytes of memory. A Polynet VMI-1 interfaced the system to the ring and a Symicron DTSX board (Symicron, 1980) provided an X.25 interface to JANET. The latter interface was a Multibus slave card, based on an Intel 8085, which implemented the physical, link, and network access levels of the CCITT X.25 (1980) specification (CCITT, 1981*a*). The system had no integral disk and was loaded from the bootserver on power-up or reset. A directly attached console was used to display logging information about calls through the gateway. Remote terminal connection was also possible and was used for debugging. No authorization, accounting, or security facilities were provided.

The arrangements for address mapping between the UNIVERSE and JANET domains were recognized to be somewhat *ad hoc*, particularly in that the onus for mapping JANET titles to addresses fell ultimately to hosts in the UNIVERSE domain; nevertheless, the mechanisms were quite satisfactory for use in an experimental situation. They were also simple to implement and achieved portability of the gateway code in the sense that there were no dependencies in the code on where it was executing. Performance of the system, while never thoroughly scrutinized, was found to be satisfactory in practice. A further point of note was that, like a number of other systems in the project having a semi-service role, logging was to a directly attached console. A more convenient arrangement would have been to allow this as an option but to use the site log for normal operation.

As a component of the overall project, the X.25 gateway was only a partial success. There were two fundamental reasons for this. Firstly, the construction of subsidiary network connections was of very much lower priority than construction of the main network. This applied not only to the gateway itself but also to the effort needed to make available in the right places some of the protocols needed to realize the stated objectives. In consequence, some of these objectives were only realized towards the very end of the project. The second problem encountered by the gateway was hardware unreliability. No effort was ever available to make a serious investigation of this. In spite of these two weaknesses, gateways were installed and used for several years at three of the sites: Cambridge, Loughborough, and RAL.

5.2.7 *Basic block tunnel*

One of the early suggestions made in the project was to use JANET to provide an alternative to the satellite segment for site interconnection in a way compatible with the UNIVERSE network architecture. Though conceived as a means for testing components before the satellite bridge was ready, in fact no effort was available until near the end of the project, in late 1983, to work on what was by definition a secondary activity.

Simply stated, the purpose of the basic block tunnel (BBT) was to provide, in co-operation with tunnels located at each of the other sites, a service equivalent in all respects but performance to that provided by the satellite bridge, using JANET as the site interconnection mechanism. From the viewpoint of a site, the tunnel presents the same service as a bridge. To replace the satellite subnetwork, the tunnels used a set of point-to-point transport service connections between sites to provide all-to-all communication between each other. This was termed the *tunnel subnetwork*. Each tunnel provides a logical bridge between the local site ring subnetwork and the tunnel subnetwork. Like the satellite subnetwork, only bridges are connected to this subnetwork. The routing algorithm used by the nameservers and bridges is the same as when a satellite bridge is used, so that the tunnel replaces the satellite bridge as the (sole) bridge nominated to convey traffic to other sites.

Provision of the tunnel subnetwork was separated from implementation of the logical bridge or tunnel. The latter assumed the existence of a ring-based transport service by which it could access the tunnel subnetwork and the availability of a mapping by which UNIVERSE site addresses could be transformed to the transport service address of the relevant tunnel. Access to the tunnel subnetwork was provided by the X.25 gateway (Section 5.2.6). The mapping was provided by use of the transformed name string facility in the nameserver. The general arrangement of components and network connections used in providing the BBT service are shown in Fig. 5.6.

The tunnel subnetwork is responsible for transporting a variety of messages between peer tunnels. Such a message may contain user data, a BRIDGEOPEN, a BRIDGESSP, an OPENACK, an SSPRPLY, a datagram, or other inter-bridge traffic such as a trace request or response. Each

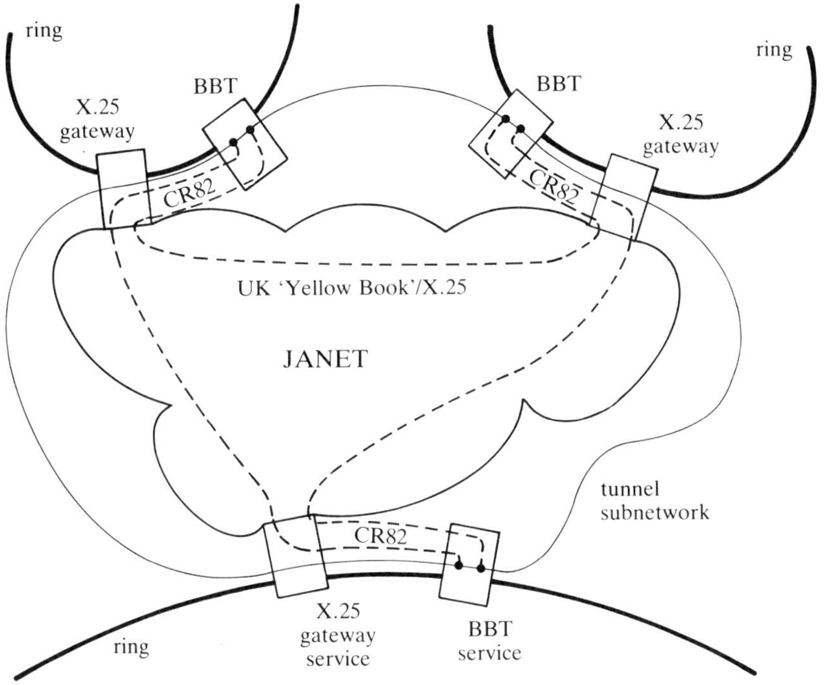

FIG. 5.6 Tunnel subnetwork components.

block entering a tunnel, though ordered with respect to any others on the same LVC, is otherwise independent of any other and must be transported expeditiously by the tunnel subnetwork to the peer tunnel. To achieve this each block is sent as a UK 'Yellow Book' transport service protocol *message* (or *data unit*), that is a set of *data* primitives followed by a *push* primitive. The latter both delimits the block and ensures that it is delivered, regardless of partially filled X.25 or byte stream packets.

To perform the mapping from the UNIVERSE site number to the UK 'Yellow Book' transport service address of the corresponding peer tunnel, the transformed name string facility of the nameserver is used in the way described in Section 5.2.6 for the X.25 gateway. At each site, entries of the form *BBT-SITE-Xnn*, for every site *nn*, are made in the name table. The UNIVERSE address returned by looking up this entry is that of the X.25 gateway service. The transformed name string contains the UK 'Yellow Book' transport service address of the peer tunnel service, which is formed of two components: the X.25 DTE address of the peer site's X.25 gateway service and the UNIVERSE name of the peer site's tunnel connection service.

Since there is no service analagous to ***reverse lookup*** by which the original name corresponding to a transformed name string may be discovered and, furthermore, the transport service calling address does not identify the UNIVERSE site of the calling tunnel, the ***explanatory text*** parameter of the transport service ***connect*** primitive was used informally for this purpose.

Much of the tunnel design and software was adapted from the ring–ring bridge software (Section 5.2.2). At the heart of the tunnel are the (slightly modified) bridge halves and support tasks of the ring-ring bridge, executing under the TRIPOS operating system. The bridge's ring device drivers are each replaced by tasks which implement 'logical devices' analagous to the ring interfaces of the bridge. In one direction, basic blocks are received from the ring and passed on to the tunnel subnetwork via a CR82 transport connection to the relevant peer tunnel. In corresponding fashion, basic blocks are received from the tunnel subnetwork via CR82 transport connections and, after routing is performed, they are forwarded onto the local ring subnetwork. For simplicity of implementation, CR82 connections are used in simplex fashion by the tunnel: between each pair of peer tunnels in active communication there is a pair of connections, one for forwarding blocks in each direction. These connections are set up automatically as the need arises; when a connection has been idle for a period, say ten minutes, it is timed out and closed.

The tunnel was designed and implemented at Cambridge to execute as a stand-alone, dedicated system on a disk-less 'large server' (see Chapter 6). It was usually loaded from the bootserver. Like the ring–ring bridge, the tunnel consulted the ***own address*** and ***read routes*** services to discover the immediate network topology. These services are consulted continually, so that if the network topology changes, the tunnel does not have to be re-booted. Three tunnels were put into service, at Cambridge, Loughborough, and RAL.

Monitoring and control were provided at a console terminal directly connected to the tunnel. Commands were available to print monitor statistics, such as histograms of block size, traffic statistics, and port usage; to display the topology perceived by the tunnel; to list information about the active ports; and to set the time-out value for transport service connections. The tunnel also supported remote monitoring by sending report datagrams to a 'site monitor' service. The reporting rate could be controlled remotely through an SSP command interface. The monitoring information could also be obtained on demand from this interface.

Generally, the tunnel achieved the aim of providing a 'plug compatible', if slow, alternative to the satellite bridge. There were four areas in which full transparency was not achieved: port capacity, buffering, end-to-end acknowledgement time-out, and broadcast.

Only 30 dynamically allocated ports were available. In practice, this was sufficient in the rather limited experimental environment. Indeed, even on the rather more actively used satellite bridge, the observed maximum number of simultaneously active ports was only 37.

The tunnel only queued an aggregate of 20 basic blocks at any one time for transmission to the tunnel subnetwork. When this limit was reached, the tunnel accepted a further two basic blocks on each allocated port before it rejected ring data. This problem is intrinsic in any bridge; in the case of the tunnel, it was exaggerated by the substantial difference in the speed of the networks on the two sides. In practice, the tunnel handled traffic which could be divided into two categories. In one case, CR82 (byte stream) protocol was in use and, in the other, the traffic consisted of sufficiently small, infrequent bursts of data as not generally to cause a problem. Both the queue limit and the port capacity could be increased if additional memory were installed.

The general question of time-outs in the UNIVERSE architecture has been discussed in Chapter 3 (Section 3.7). Only the information that a service is 'local' or 'remote' is available to a client to assist in gauging network delay when determining a suitable time-out value for transactions or connection establishment. The inadequacy of this information in general was clearly demonstrated by use of the tunnel system, since the X.25-based tunnel subnetwork could on occasion introduce a substantially greater network delay than the satellite subnetwork. This delay was even more serious when using the byte stream protocol, which was designed for a high-speed LAN environment and had a window of one. The time-out value associated with window acknowledgement was sufficiently low as to reduce throughput to zero on occasions when JANET was loaded with traffic. As an unsatisfactory palliative, the BSP acknowledgement time-out was increased but the fundamental problem remains.

In summary, the limited experience gained by three of the participating sites in the use of the tunnel system made it clear that it would have been valuable to have had such a facility early in the project to assist in software development. Full emulation of the satellite subnetwork would have required provision of a broadcast service. While this would no doubt have been useful, it was never planned as part of the tunnel system. It should be emphasized that the BBT concept was in no sense viewed as fundamental from a network architecture standpoint within the UNIVERSE project. However, certain aspects of its implementation, notably the separation in handling traffic in each direction, led to some speculation about its possible usefulness as a basis for developing a bridge for a 'star' subnetwork, based on a number of receive-only earth stations, a central 'hub' transmit station, and a number of terrestrial lines linking the latter to the former. Such a configuration might be interesting for data distribution applications using a relatively high-speed space segment with low-speed terrestrial lines. In the context of UNIVERSE this idea remained entirely speculative.

5.3 Commentary

This chapter has concentrated on describing the particular realization of the UNIVERSE network architecture which was constructed and operated for experimental purposes, from experimental equipment during the lifetime of the project. Considerable effort was expended in this aspect of the endeavour and it is appropriate to record a number of overall observations assessing the successes and shortcomings of the implementation.

5.3.1 *Operational issues*

As might be expected of a network constructed especially for experimental purposes in a short time, availability of the complete network in an operational state was low. This was primarily due to the use of a considerable number of experimental pieces of equipment, each of which was itself in a largely experimental state: relatively few of the network components were engineered to operational product status and consequently required substantial expenditure of expert effort to maintain in operation. Nevertheless, a sufficient proportion of the network was available often enough to permit a substantial amount of experimentation to be carried out and for three public demonstrations to be mounted.

In an operational network, considerable attention needs to be devoted to the engineered provision of integral monitoring and diagnostic facilities. This area was never seriously planned or attempted in UNIVERSE, partly through lack of time and resources to develop the network beyond its initial, experimental phase and partly because the development of appropriate tools and techniques was itself regarded as needing study. Some effort was devoted to an investigation of the latter and is the subject of Section 7.2 of Chapter 7.

A useful, simple device which came to be incorporated in a number of network components and application systems was the so-called *code-id* service. This was an SSP service which returned an integer, representing a version number, and a text string. It was conceived as a way of discovering if a system was 'minimally alive' and, if so, what version of code it was executing. In a system supplying a variety of services, the concept could usefully be applied to each service. In terms of network operational assistance, the positive acknowledgement aspect of this service was not only reassuring but also established the existence of an operational route in the network. A generalization of this service was provided by the LDC, which gave status information about the satellite segment. These sorts of service are simple to implement and provide cheap, basic checks of system and equipment functionality.

The arrangements for booting the various network elements from a central service on each site were in principle quite adequate for experimental purposes and were in practice marred only by the unreliability exhibited by the particular equipment chosen for this role. This emphasized the inherent weakness in a strategy which placed reliance on a single system in a central role. In fact the booting procedures implemented in several of the larger systems (all the 'large servers', for example) took account of this aspect by trying to boot from an alternative, 'reserve' bootservice in case of failure of the primary service. Although not studied any further in the project, for an operational system, it seems likely that still more flexibility in this area would be desirable: more thought would be needed about how best to provide this. The question of how to keep the contents of the bootservers consistent throughout the network, while recognized as a problem, was not addressed: moreover, the scale of the project was small enough for manual procedures to be adequate for the duration of the project. For a research and development project, a more serious shortcoming was the lack of a complete repertoire of network access facilities on the bootserver system. This was the more apparent because of the presence of other systems better provided in this respect.

Generally, systems such as the nameserver, bootserver, site log, etc. were always left switched on. However, interdependencies among these basic services were such that the order in which systems and processes were started up was important. A thorough analysis of this area would be necessary for an operational system.

5.3.2 *Use of JANET*

It has already been emphasized that the availability of this network to the project aided considerably in both the development of some network components and in software sharing. Had the X.25 gateway been both more reliable and more generally available earlier in the project, it is likely that JANET would have seen still greater use. Two aspects of this can be identified. The first was a question of protocol availability. The Cambridge TRIPOS system had associated with it a set of a well-established network utilities for file transfer and remote terminal access. These utilities made use of protocols which were different from those in use on JANET. However, they worked well enough over the satellite for infrastructure use and their presence effectively prevented much work on duplicating the facilities they provided by developing parallel utilities using protocols which would allow interworking with systems on JANET. Some work was done in this area but it was never promulgated very widely in the project. In terms of obtaining experimental results, the issue was probably marginal in UNIVERSE. For a larger project of longer duration, it would probably be important to attend to this area.

The other topic was that of the BBT. Only three sites had experience of using this and that was really after the main part of the project had terminated with the demise of OTS access. Nevertheless, more than a year's use of this facility served to demonstrate that it would have been exceedingly valuable had it been available over a year earlier, particularly in speeding development of some of the distributed systems experiments and enabling them to reach a more mature state before the end of the project.

5.3.3 *Name table management*

This was another area where sufficient was done to allow the project to proceed but development was not pursued to the point where it could be recommended for a product system. Two types of desirable development can be identified here: improvements in operational usability and in fundamental mechanisms.

Name tables were arranged so that within a single domain there could be a single master name table residing in the (master) name table read service. Updating was performed in two stages: editing a text file, assembling it to produce a new master name table, and copying or transferring it to the name table read service, followed by stimulating all the nameservers in a domain to reload their name tables. Since a nameserver reloaded the whole of its table (or none at all), the problem of internal inconsistencies was largely obviated. No such guarantee obtained between different nameservers: without manual intervention it was possible for a nameserver to remain indefinitely inconsistent with respect to other nameservers in the same domain. It would have been a matter of straightforward development to introduce more structured assemblers (as was eventually done) with more internal checking. Some considerable improvements could also have been made in the operator interface to the whole name table update procedure. More substantial would have been the development of an interactive means of update of individual entries. Notwithstanding that the effort to do this in the project would have been not inconsiderable, no new principles would be involved.

More fundamental from a network architecture point of view is that the very property which makes the concept of a master table so attractive, namely its singularity, also carries the seeds of its weakness because it represents a single point of failure. In the project, there was no provision for a backup machine to hold the master table and provide the name table read service (which is amenable of remedy), for a nameserver to obtain its table from more than one service, or for a nameserver to use any alternative route to reach the name table read service. The latter is a single aspect of the fundamental fixed routing strategy adopted in UNIVERSE; this is an important issue which is commented upon in Chapter 11.

5.3.4 *Conclusion*

It will be observed that while numerous areas have been identified where improvements would be necessary for an operational network or would have been helpful to the project, none of these prevented experimental results being obtained. The single most significant hindrance was the very experimental nature of some of the equipment: this was recognized from the outset and an operational network could not be expected to operate without properly engineered components.

Although there were many ways in which the network implementation could have been improved, it has to be remembered that the planning for the whole project and the implementation of the network were undertaken in two years, from the time of initial approval to the first public demonstration. That this was achieved was due not only to the substantial effort expended by those charged with the network construction and development but also to the basic simplicity of the overall network architecture.

6. Experimental programme

6.1 Outline and organization

The account of the project has so far dealt almost exclusively with the architecture and construction of the network itself. It is implicit that networks are a means to an end and not an end in themselves. To demonstrate the working of a network and to explore its capabilities and limitations it is necessary to exercise it. To this end a programme of experiments was developed by the project. This allowed a wide variety of traffic to be tried out over the network. It also acted as a convenient vehicle for organizing the varied interests of the project participants.

The means for deciding and organizing the experiments was quite straightforward: participants declared their interests in various topics and invited collaborators. Once approved by the Principal Investigators' Committee as the basis of an experiment, it was assigned to one of the working groups for technical discussion and progress. The only exceptions to this were those experiments which were essentially solo efforts by one of the participating organizations: in such cases the organization and management of that experiment was left under the control of that organization, though one of the working groups would still be designated to act as a forum in case technical discussions were needed. A complete list of experiments undertaken by the project is given below, categorized by subject.

I Network monitoring

II File access and transfer

1. Remote file access protocol
2. Remote use of fileserver
3. Broadcast file transfer
4. File distribution system

III Real-time traffic

1. Voice
2. Image transfer
3. Image compression studies

IV Security

1. Authentication
2. Encryption

V Special terminal application protocols

1. IBM 3270 full-screen access to VM CMS
2. Videotex
3. Teletex

VI Distributed integrated programming support environment

VII Reliability issues in LANs

1. Design of a reconfigurable token ring
2. Reliable topologies for ring-based LANs

In any network activity, the question of monitoring network performance and activity arises. The subject is a very broad one: it can cover questions of reliability, both of individual components and subsystems; collection of traffic statistics, classified by links, routes, switching nodes, subnets, hosts, protocols, services, time of day, etc.; also measurements of performance along particular paths, across subnets, between hosts, either during periods of congestion or with no other traffic present. In any network these sorts of measurements represent the behaviour of the system. In an experimental network, they are needed to determine the areas in which the design is successful or needs improving. In a more service-oriented environment, such figures provide the fundamental input needed for operational control, fault diagnosis, charging, and capacity planning. In UNIVERSE, this topic was attacked on several fronts by UCL. However, the need for specific measurements in particular areas led to other participants making certain complementary measurements. Likewise, in presenting the results it has seemed appropriate to present some of these in the context of the component or experiment to which they directly relate. The more general activity is described in Chapter 7.

In a network purporting to provide high-speed communication, it is clearly very important to examine in some depth the degree to which the network succeeds in this respect. In any network with a satellite segment it is obvious that in respect of one measure of 'high speed', namely transmission delay, such a network may be at some disadvantage compared to others. In terms of rapid bulk data transfer, however, such a network ought to be able to provide satisfactory high-speed performance. The next two topics, file access and transfer and real-time traffic, represent the major fraction of the effort put into experiments by the project. They also represent the major focus for the effort that went into determining the degree to which the network could be regarded as high performance. Most of file access and transfer is covered in Chapter 7; the exceptions are remote file access protocol and transfer performance measurements, which are covered in Chapter 9. Real-time traffic is the subject of Chapter 8.

Use of a broadcast satellite as a principal segment of a network raises the question of secure communication in a particularly obvious fashion. However, the subject is one of very broad application and in consequence two of the commercial participants together with Cambridge University undertook some experiments in this area. The authentication system is an example of a distributed system and is described in Chapter 7. Encryption is covered in Chapter 9.

The availability of an experimental network with claimed ability to carry any type of traffic or protocol led several participants to seize the opportunity to try implementations of a number of specialized application protocols, listed above under section V. These experiments are described in Chapter 9.

A recurring topic in the study of computing is software engineering, the goal of which is to provide improved techniques for the specification, costing, construction, and maintenance of software. The subject is one never far from general attention since inadequacies of current techniques are generally held to account for many of the unsatisfactory aspects of software production. One area of investigation is that of improving the engineering environment for the software development process. At the time of the project, this concept was referred to as an 'integrated programming support environment' and was particularly stimulated by the need for such an environment suitable for the building and testing of real-time applications programmed in the ADA language. In this context Logica conceived the idea of constructing such an environment based on distributed systems design principles (VI above). The impetus for this arose from their experience that in building such systems it was not infrequently the case that the various resources needed for development and testing were not all to be found in one system nor even at one location; moreover, moving the resources was not always practicable. This experiment was a substantial and ambitious undertaking; it was also an interesting application of ideas arising from distributed computing research and is described in Chapter 7.

Two activities, both stimulated by the project but undertaken somewhat independently of it, are brought together under topic VII. GEC undertook the construction of a token ring, with issues of reliability very much in mind. The automated factory was taken as an example of an environment where such reliability is needed, on both safety and economic grounds.

Loughborough undertook a study of reliability issues in ring-based LANs and how unreliability in various of the components might be alleviated by more complex connection topologies. Both of these topics are taken up in Chapter 9.

6.2 Support infrastructure for experiments

For a range of the experiments constituting the programme for testing the properties of the network, the emphasis was on demonstrating particular styles of use or making specialized measurements, not on demonstrating that the architecture was 'open' in the sense of international standard open system interconnection. Whilst in the latter context it would be appropriate to have a number of different systems, in the case of the former it is a serious disadvantage. Not only does the use of a variety of systems imply considerable expenditure of nugatory effort in reprogramming applications or network measurement programs to run on different systems, it can also make some types of network measurement more difficult to perform. In retrospect, it is the view of a number of the participants, shared by the authors, that even more emphasis should have been placed on the uniformity of programming and experimental infrastructure.

Simple though this may sound in principle, it has a number of implications for the program development infrastructure; for only some of these was a satisfactory approach devised in the project. In developing software for a machine, it is necessary to have not only an execution environment for that machine which is suitable to the task in hand but also a programming environment suitable for compiling and loading the software into the target machine. In some cases it may also be necessary to have available special test execution environments capable of giving more run-time test and diagnostic output than is usually easily available by simply loading and executing. At this point a very significant problem becomes apparent in a collaboration such as UNIVERSE. Most research and development groups will, over a period of time, have evolved a programming environment oriented to their own particular needs and practices. This environment will inevitably be strongly tied to the languages and operating systems with which a group habitually works. It is also more than likely that a set of supporting software tools will have been built up over a period of time; it may also be used by a larger set of people than any one particular group. Support for this infrastructure will be spread among various parts of an organization and some may well originate outside it altogether. Thus the programming environment for a group is likely to be deeply embedded and to represent a considerable investment of resources. Any change is likely to be very costly and time consuming. It is also very unlikely in any collaboration akin to UNIVERSE that much commonality will be found in place. It is after all an indigenous feature of research and development groups that useful programs and techniques resulting from previous successful investigations are incorporated into the environment.

UNIVERSE has no solution to offer to this problem in general. Here we confine ourselves to noting that the problem existed throughout the life of the project and was never satisfactorily resolved. Moreover, it is to be anticipated that the situation will occur generally in collaborative projects and will need to be taken into account at the outset and incorporated into the cost of collaboration when estimating the budget. It was the experience of the project that anything which can reasonably be done in this cause will well repay the investment.

In UNIVERSE there were really two main requirements relevant to this discussion. There were system items which needed to be constructed and then replicated at each site. Examples were the nameservers and bootservers described in Chapter 5. For most purposes these could be regarded as 'black boxes' by all but those responsible for their construction. Since the construction of such items was assigned to one organization, the approach taken was to leave, where possible, the choice of both hardware and software to the responsible organization. The other requirement was to have available at every site the wherewithal for distributed systems experiments to be performed. Some thought was given to the possibility of having hardware in place into which it would be possible to load alternative operating systems. Even here substantial difficulties exist. Most of the systems used for distributed computing had a Winchester disk. Unless the storage on such disks is regarded as completely volatile, it will be arranged according to some particular operating system's filestore. That operating system is then the only one which can be loaded when the disk is required. Nevertheless, it was recognized that for distributed systems work, even if the target system was not ideal, it would still be better only to have to program a process once.

For UNIVERSE it had been hoped that a commercial system could be found that would be a suitable basis both as a bootserver and for distributed computing. While it was never intended that the bootserving role should be provided other than on a dedicated system, nevertheless it would have been very convenient for the base systems to have been identical. Unfortunately this plan failed because the system chosen for the role of bootserver, after a necessarily short period of assessment, was later found to be unsatisfactory in a number of respects and a different system was chosen for the distributed systems experiments. This latter system was commonly referred to as the 'large server' and is described below, in Section 6.4.

6.3 Cambridge distributed computing system

It is necessary for an appreciation of some parts of the experimental programme to give a short description of the Cambridge distributed computing system. A complete description may be found in Needham and Herbert (1982). The overall objective could be described as the construction of a system capable of providing users with services similar to those of a conventional time-sharing system. However, the system is built out of a set of distributed components connected by a LAN. The principal difference, from the user's viewpoint, is that instead of being given access to some fluctuating fraction of a central processor and its central memory, the user is allocated a private processor and memory for the duration of a session. All other system resources needed by a user are provided on a shared basis by other (distributed) components of the system. The most obvious examples are filing services and printer support. The system may be considered to be constructed out of a number of levels (see Table 6.1), the topmost level (4) being the single-user processors which it is the purpose of the system to support on a personally and dynamically allocated basis. These processors were generally Computer Automation LSI 4s or locally produced, Motorola 68000 based systems. The LSI 4s came in a range of CPU speeds; they have 16-bit address architecture, 16-bit words, and 64 Kword of memory. The 68000 systems generally had 512 Kbyte of memory. In both cases, the only external connection to these processors was the Cambridge ring.

To make these systems useful, means must be provided to allocate and load them with an operating system; a filing system must be provided; access from the user's terminal needs to be arranged; and hardcopy output will undoubtedly be required. These functions may be considered the essential ones provided by the next level (3) of the system. Starting with the user's terminal, access is provided by means of terminal concentrators, each handling eight keyboard and screen terminals. When not connected to anything else, by default these connect the user to a service called the session manager. Demands are made of this service to set up a session with a particular processor, or class of processor, and for it to be loaded with a particular operating system, usually one of a range of TRIPOS (Richards *et al.*, 1979) configurations. Processor bank resource management is the responsibility of the resource manager service. Actual loading of 68000 systems is performed by an entity known as the ancilla (Shepherd, 1981), which maintains a private filestore of load images on the fileserver (see below) and uses a special protocol for booting 68000s. Once a processor has been allocated and loaded, the system is provided with a shared filestore identifier appropriate to the operating system loaded: usually the TRIPOS shared filestore. The session manager then changes the terminal connection so that the user is no longer (by default) in communication with it

TABLE 6.1 Levels in distributed system

Level	*Element*
4	Single-user processor and memory LSI 4s, 68000s
3	Terminal concentrator (8 VDUs) Session manager Resource manager Authentication server Fileserver Filing machine Print server
2	Nameserver Bootserver (and ancilla) Date and time server Site log Error log
1	Cambridge ring LAN Monitor station Repeaters and associated stations Slave power supplies Ring error logger (station 0)

but is instead connected directly to the newly provided processor. The final activity is for the user to provide an identifier and password which are verified by use of the authentication service, about which more will be said later in describing the authentication experiment (Chapter 7). The user's system is then connected to the relevant user node in the filing system.

The TRIPOS shared filestore service is provided by co-operation between two services with quite distinct functions. Individual clients of the TRIPOS file service communicate with a service known as the filing machine. This is responsible for implementing a shared, tree-structured directory based filestore. It maintains a shared cache of commonly accesssed parts of the filestore. Access by client systems occurs at the granularity of filestore blocks. The filing machine has no disk backing storage of its own: instead, it makes use of a fileserver. The latter has the function of providing identifiable storage of sequences of bytes. This storage is managed, garbage collected, and protected from corruption through hardware malfunction by regular backing up to removable media. The fileserver provides no higher level of storage structure abstraction.

There is a variety of output services attached to the system which are more or less application-specific. However, there are also general-purpose printing services, generally provided by laser printers driven by dedicated 68000 systems. The basic printing service has no spool storage associated with it; instead, it simply waits to be claimed by a client which then prints its document and releases it. Requests from other clients are ignored while it is busy.

Level 2 of the distributed system provides a number of infrastructure services upon which those already mentioned rely for their general operation. Principal among these are the nameserver and bootserver. The role of the former is described in detail as part of the UNIVERSE architecture (Chapter 3); its implementation is described in Chapter 5. In the current context it is simply the sole repository for the binding of names of machines and services to network addresses.

A number of services have been mentioned which form an integral part of the implementation of the distributed computing system. These services are mostly implemented on dedicated processors and themselves require loading at switchon or after a software malfunction. It is the purpose of the bootserver to provide this service. A more complete description of this service is given in Chapter 5 since it was one of the services replicated at each site participating in UNIVERSE.

Several other miscellaneous services are also considered to belong to level 2. One of these is a date and time service, which provides this information on demand on the basis of the continuous standard time signal available by radio from Daventry, near Rugby, in Warwickshire, England. A general logging service is also available. It is generally intended for the logging of exceptional or erroneous events.

The lowest level (1) of the system is considered to be the ring itself, by which all the other components communicate. Apart from the essential monitor station and the repeaters and stations by which systems are attached, this also includes slave power supplies and facilities at station address zero for the reception of ring logging of errors and diagnostics.

6.4 Large and small servers

A fundamental thread throughout the project was that of distributed systems. As has already been described, some of the architecture and much of the implementation of the network was based on such principles. Both here and in the relevant experiments the attitude was that, for the purposes of what was in any case an experimental project, one should not seek to redevelop equipment or techniques if one of the participants already had in operation suitable software or hardware. Since Cambridge had had a distributed computing system in operation for some years, this led to replication of a number of items from that system.

6.4.1 *Large servers*

The need for a set of identical systems to reside at each site in order to enable distributed computing experiments to be carried out with the minimum of programming effort has already been explained. There was considerable debate within the project about what systems should be chosen to fulfil this role, during which the following constraints and criteria emerged. (It should be borne in mind that the decisions on these systems were being made in 1983. Considerable improvements have been made in price and performance since then.)

(a) The system had to be relatively inexpensive. Specifically, the cost needed to be in the range £5k to £10k: the upper limit was determined by a combination of budget and need to have at least one such system at each site. The lower limit was an empirical one, below which it was inconceivable that the system would satisfy enough of the necessary criteria to be of any use. It was also recognized that it would be almost essential to have more than one such system at some of the sites: this was an additional inducement to strive for a system at the lower end of the price range.

(b) Since several of the experiments to be carried out with the aid of these systems involved either the collection of data or the provision of staged file transfer, disk backing storage was essential.

(c) The system required sufficient memory (about 512 Kbyte) to support a disk-based operating system and real-time applications programming in a higher level language than assembler. Memory addressing was required to be unsegmented.

(d) It was required to interface the system to the Cambridge ring and to be able to boot it via the ring. Good performance of the ring interface was necessary.

(e) Experience with the system used as the bootserver had already indicated that whatever system was chosen, expert knowledge on all its aspects, both hardware and software, should be easily available to the project.

The price consideration in item (a) weighed in favour of a system based on a microcomputer; item (c) effectively implied use of the Motorola 68000 processor; and item (d) reduced the choice to a system based on the Intel Multibus or to copying one then in prototype developed by the Computer Laboratory at Cambridge and Acorn Computers Ltd. Item (e) and the booting requirement resulted in the choice of the latter. That the price was also at the lower end of the range, allowing 12 systems to be purchased, 10 with disk, was also a considerable advantage.

These large server systems were constructed from double-height, extended-depth Eurocards, housed in a 19-inch crate with integral power supply. The main backplane was proprietary and served to interconnect a 68000 CPU card, a 512 Kbyte memory card, and a 6809-based intelligent DMA controller for the ring. An interface to a 1 MHz bus was used to support a cheap 10-Mbyte Winchester disk and controller, and also a simple keyboard and visual display unit (VDU) terminal interface. The 6809 board had 64 Kbyte of memory, of which 4 Kbyte was ROM containing basic block protocol support for communication over the ring, the booting program, and a remote debugging interface which allowed another host on the ring to perform the lowest level of software debugging on the 68000. The system had an integral connection to the Cambridge ring, in the form of SEEL station and repeater cards mounted in the crate and connected to the 6809 ring interface.

The operating system used on these systems was TRIPOS (Richards *et al.*, 1979), a research operating system developed at Cambridge some years ago. The system scheduler is priority based. The entities which are scheduled for execution in this system are called tasks and each has a single work queue of messages. Message-passing between tasks is the only form of task communication which the system supports. Since the system uses a single, unprotected address space, other forms of communication between consenting tasks are not prevented but they will have no effect on task scheduling. Tasks are only suspended because they perform a voluntary wait for another message or because a hardware interrupt occurs. Peripheral devices are handled by drivers which also have work queues of messages. These are entered whenever an appropriate interrupt occurs or a packet is placed on an empty work queue. Exit from a driver is by way of the scheduler since the driver may have sent a message packet back to a task and the highest priority task now ready to execute may no longer be the same as when the driver was entered.

The kernel of this system (the part requiring intimate knowledge of the scheduler, providing space management, system initialization, task manipulation, including creation and deletion, and debugging primitives) is written in assembler. Device drivers and some library routines are also in assembler. The rest of the system is constructed in BCPL (Richards and Whitby-Strevens, 1979) and consists of a comprehensive version of the BCPL library and a set of system tasks. The usual set of tasks available on a large server consisted of a console task, debug task, file handler, command line interpreter (a very primitive user 'shell'), ring handler, and ring services task.

In the context of modern application program development environments, TRIPOS should not be considered a contender. Its usefulness in the present context lay in two areas. Its scheduler is both simple and fast; its message-passing is also achieved with reasonable celerity, particularly as messages are not copied. Context switching to handle interrupts is handled equally expeditiously. Thus the system can reasonably claim to be a real-time operating system. The other feature of note is that it is a very simple matter to turn a user program (with appropriate functionality) into a system service task. This is particularly the case for services provided via the ring. This ease of system task development and integration is seldom found in other, especially commercially available, operating systems.

Since the TRIPOS operating system and systems similar to the large servers had been in operation for some years on the ring at Cambridge, these systems came with a number of programs for ring testing and communication already incorporated. Examples included checking on every station on the ring, directing an SSP to an arbitrary station and port, client and server terminal access, and file transfer facilties. These last enabled software maintenance of these systems to be performed entirely over the UNIVERSE network. An infrastructure role in which these systems turned out to be useful was in assisting with the maintenance of nameserver name tables.

6.4.2 *Small servers*

Typically, a small, low-level or dedicated service in the Cambridge distributed system is built using a Z80-based, single-board computer interfaced to the ring. Such services include error loggers, a date and time server, terminal concentrators, and nameservers. It was anticipated that, for a number of small tasks, such as traffic generation, packet reflecting, and the like, it would be convenient to have a number of such systems available throughout the UNIVERSE network. Here a brief outline is given of the hardware for these systems. Descriptions of systems developed specifically for UNIVERSE, such as the telephone controller (Chapter 8), are given elsewhere, in the context of their use.

The Z80 small server was constructed on a double height, extended depth Eurocard and had 64 Kbyte of memory. 1 Kbyte of this was EPROM containing basic block support for ring communication, a ring booting program, and support for remote debugging via the ring of programs in Z80 RAM. Small four-position backplanes were made: two positions were occupied by SEEL ring repeater and station cards and one by the Z80 card; the fourth position was available for a peripheral card. Three of these systems fitted into a 19-inch crate with power supply.

Nameservers for each ring in the project were constructed using these small servers. For these the Z80 CPU card was configured for 8 Kbyte of EPROM, to hold the nameserver code and the small default name table, described in Section 5.2.4.

6.5 Subsidiary developments

One of the necessary and time-consuming tasks undertaken as part of the project was the development of interfaces to the Cambridge ring for several types of computer. This included both hardware and software development and the replication of the hardware. Probably the most ubiquitous interface component in the project was the Polynet VMI-1, developed and manufactured by Logica VTS. This was a single board for use in Multibus-based equipment and provided reception and transmission of basic blocks under DMA control. Control of the device was through a shared data structure located in memory situated elsewhere on the Multibus. The board was based around an Intel 8086 processor and also included additional logic to assist in receiving minipackets fast enough to keep up with the full point-to-point data rates on the ring. The principle weakness of the VMI-1 was that it was half-duplex at the basic block level. This board was used in a number of Multibus systems throughout the project to provide a direct interface to the ring. Examples included the bootserver, 8086 TRIPOS systems at RAL, the X.25 gateway, the framestores developed at Loughborough, and MDS development systems used at GEC.

A more indirect way of implementing an interface, suitable for larger systems, is to map a computer manufacturer's proprietary bus or other interface to a more standard bus and then apply a known interfacing solution to the latter. This method was employed successfully for three different manufacturers' machines and clearly demonstrated the value of using such standards to reduce development risks and lead times. However, it should be noted that it does not necessarily lead to the most economical solution and will generally only be appropriate for larger systems.

This technique was successfully used in conjunction with the VMI-1 for both GEC and Prime systems. In each case, boards were developed to map parts of the main memory of the minicomputer into Multibus address space, so that memory access was shared between systems on the Multibus and the host computer. A means of passing interrupts in both directions between Multibus and the minicomputer host was also provided.

In the case of the GEC 4000 system, RAL developed a Multibus board which performed the memory mapping; interrupt signalling was handled by a standard 8085 single-board computer. Basic software to drive this interface was also written by RAL. The most important system to use this interface was the satellite bridge LDC. Host systems at RAL, GEC, and BTRL also used it, the latter to provide Prestel database access to UNIVERSE. More comprehensive driver software for OS4000 was developed by GEC, who also implemented basic support for the CR82 byte stream transport protocol (Larmouth, 1982). The latter was used to support both

TS29 terminal access (PSS User Forum, 1981) and the remote file access protocol developed as one of the experiments described in Chapter 9.

For the Prime range of systems, a special Prime board was developed under contract to the project by Cambridge Digital Systems (CDS). This board handled both memory mapping and interrupt signalling and was connected to a simple line-driver Multibus card by a length of ribbon cable. CDS provided a simple driver for inclusion in Primos; basic SSP library facilities were developed by the project. The interface was used by the computer centre at Loughborough and by Logica to make available the Rapport database system for the USE experiment (Chapter 7).

An analagous approach was taken to providing an interface to the Cambridge ring for the IBM 3081 then at RAL. An 8911 channel interface was purchased from Auscom Inc. of Austin, Texas, USA. Originally intended as a means of interfacing graphics devices to IBM System/370 block multiplexer channels, this equipment was based on a DEC LSI 11/23 and its associated Qbus. A special, microprogrammed board provided support for the block multiplexer interface. Another Logica VTS product provided a simple minipacket interface to the Cambridge ring with the aid of a Polynet node.

Software was written for the 8911 to provide support for the basic block protocol on the ring. The interface on the IBM side appeared to the host as a device controller. Support was provided for up to four device subaddresses, so that up to four virtual machines in the IBM host could have multiplexed access (at basic block level) to what appeared to each to be its own ring device. Since there was only one real device, these virtual machines could only be addressed from the ring by judicious use of port numbers. As might perhaps have been anticipated, this interface arrangement was not capable of high performance: the processing overhead for each minpacket was quite high and, using maximum sized basic blocks, the LSI 11/23 was capable of achieving a thoughput of about 100 kbit/s. The main use of this device was in the remote full-screen access experiment (Chapter 9), though support was also developed for simple file transfer.

The final system for which the project commissioned the development of a ring interface was the BBC microcomputer. This 6502-based machine has a 1 MHz bus available externally and Acorn were contracted to develop a simple programmed interface between this bus and a SEEL ring station. The arrangement was packaged so that the interface fitted into the small server crate described in Section 6.4.2 in place of a Z80 CPU card. Connection to the SEEL ring station and repeater was via the backplane; a socket mounted on the front edge of the board provided for a ribbon cable connection to the BBC microcomputer bus.

The very nature of this ring connection meant that, in contrast to the three already described, it was more expensive than the host. While this might be acceptable for experimental purposes, it underlines the well-known fact that for personal computer or small workstation attachment to a LAN, the only economically feasible arrangement lies in the use of special-purpose chips, mounted either on the main computer board or provided as a single, small add-on board.

A feature of the BBC microcomputer design (called 'sideways paged ROM') was the ability to select under program control which of 16 ROMs was currently accessed by a fixed-address, 16-Kbyte partition of the total 64-Kbyte memory address space of the 6502. This was generally used to provide support for language systems or utility packages. Cambridge provided basic block protocol support for the ring interface as a ROM utility package, accessed effectively as an additional set of system calls. Apart from providing a few simple ring exercising programs, this microcomputer with its ring interface was used extensively in the demonstrations mounted by the project (Chapter 10). Stimulated by this, some considerable effort was expended in tuning software for this interface and, using 2 Kbyte basic blocks, transfer rates in excess of 100 kbit/s were achieved.

Mention has already been made of the use of byte stream protocol (BSP) over the UNIVERSE network. In Chapters 3 and 5 its use is described in connection with the X.25 gateway and the BBT. The original BSP was developed at the Computer Laboratory at Cambridge University for use over a single Cambridge ring. It is a duplex protocol constructed with a window of one, giving error-free, flow-controlled data transfer over the basic block protocol. It is initiated by an OPEN/OPENACK exchange of precisely the form used in the project to set up an LVC. The service enables streams of bytes to be transmitted and received simultaneously. It is also possible to force transmission at any desired point within the data stream. Details may be found in the appendix to Needham and Herbert (1982). It operated unchanged over the UNIVERSE network, albeit with very limited performance across the satellite segment. Considerable standardization effort was expended within the UK on this protocol; this included some effort from the project to extend the window to values greater than one to improve performance over the satellite. This was found to be quite difficult and was not pursued. Throughout this book the term CR82 is used to refer indiscriminately to the various standardized forms of BSP. (*Note:* Following the definition of the UK transport service (PSS User Forum, 1980) used in JANET, a derivative protocol, known as TSBSP (Dallas, 1981), was developed at the University of Kent, which provided UK transport service over a single ring. Subsequently, as part of the effort to define a UK interim standard version of the Cambridge ring and its protocols, a further derivative of the protocol was defined, known as CR82 (Larmouth, 1982; see also Sharpe and Cash, 1982). Further work has since resulted in the production of an international standard (ISO, 1988) for the Cambridge ring.)

In UNIVERSE, BSP was convenient for general use as a reliable protocol since implementations already existed; for purposes where throughput was not of the essence, it was adequate if slow. At Cambridge a further development (known as AuthBSP) was made to this protocol by the addition of authentication, using the scheme described in Chapter 7. Utilities which used BSP, AuthBSP, and CR82 for file transfer and remote terminal access were available in TRIPOS. Support for some of these was also provided by GEC in OS4000 (see above). Within the experimental programme, USE (Chapter 7), teletex, remote file access, and encryption (Chapter 9) made incidental use of these protocols.

Almost all of the infrastructure development described here was used by more than one experiment, site, or system. The exception was the IBM 3081 system interface, which only existed at RAL and was only used in a single experiment. Examples of rather more peripheral infrastructure included arrangements for remote debugging, particularly of small servers; a variety of protocols (and corresponding implementations) for remote terminal access and file transfer; and a mechanism for two-stage booting of one of the systems (the remote filing machine, Chapter 7) via the satellite, from one site to another. All of these were of importance in some way to the support of the experimental programme. However, the support they provided was of an indirect nature, as for example in assisting in remote system support for the large servers, rather than being an integral part of an experiment.

In this chapter, a general introduction to the experimental programme used to evaluate the network has been given and the more pertinent aspects of the experimental infrastructure have also been described. It was known at the outset of the project that in order to support any of the experiments, the 'raw' facilities of the network would require an overlay of infrastructure. This was especially so for the more ambitious of the distributed systems experiments. This infrastructure continued to develop and expand in richness until about the middle of 1983, six months before OTS access was due to end. A good deal of the project effort was naturally expended on this area, especially for those new systems or subsystems developed within the project. Although some experimentation was begun in 1982, the majority was undertaken in the final year of OTS access, in 1983. A recurring theme in the following chapters is that by the end of 1983 about half of the experiments had, in the course of their investigations, not only shown that a certain approach was feasible and obtained some preliminary results, but had also revealed further worthwhile directions for investigation. It is of course in the nature of research that this should happen and it is the assessment of the authors that a programme of experimentation for a further year within the collaboration, using the network and infrastructure already in place, was generated as a result of the experimental application of the network. Had it been at all possible, it would have been a worthwhile use of the resources to have extended the period of experiment, to provide a correspondingly greater yield in terms of research results for the investment of effort.

7. Distributed systems

7.1 Introduction: distributed system mechanisms

Collected together in this chapter are descriptions of those experiments which, while not otherwise particularly related to each other, did have in common that they were examples of distributed systems. Among the results of each experiment may be counted assessments of both the appropriateness of the distributed approach to the design of the particular system and the degree of its success in that context. Taken together, these systems span quite a wide range of distributed computing activities and allow some overall assessment to be made of the suitability of the network as the basis upon which to implement wide area distributed systems.

In Chapter 1, a broad description is given of the nature of the distributed systems under consideration in the context of UNIVERSE. The general approach is to separate the major component activities of a system and realize these on individual computers connected by a network. Among the potential advantages of such an approach are the use of replication for increased capacity and reliability in particular components. The LANs used for this purpose have properties which lie somewhere between those of a traditional wide area network and those of a backplane bus system. Not only may this be said of the basic hardware mechanisms employed, it is also true of the types of protocol which have been developed for use in this context. The parts of a distributed system generally communicate by means of relatively short messages used to request action, to convey results, or to achieve synchronization. Somewhat longer messages are used to convey bursts of data. The general system design philosophy is akin to that used in single-processor operating systems based on message passing. Since it is important that such messages are passed with the minimum of delay, connection-oriented communication, which usually requires a separate setup phase, is typically not used. The obvious vehicle for such messages is the datagram, particularly since, as was pointed out in Chapter 2, in the case of a solitary LAN this represents a very small addition to the fundamental means of passing packets across the network.

The conventional datagram begins to exhibit some disadvantage when LANs are interconnected. Precisely the property which is advantageous with respect to being amenable of flexible and robust routing is also responsible for contributing additional processing overhead within the interconnecting network bridges. On the premise that in a high-speed network the CPU power of the bridges is likely to be a limiting factor, this is particularly unfortunate.

An especially common form of activity in distributed systems interactions is that of making a request and receiving a reply, where either or both parts of the transaction may be short enough to be contained within a single network packet. The single shot protocol (SSP) was already in use for this purpose at Cambridge and was extended across the network as one of the UNIVERSE network access protocols. The mechanism adopted to achieve this required the network elements to perform some setup processing in order to transmit the request but no further such processing is needed to pass back the reply. This saving is achieved at the cost of sacrificing any flexibility in regard to the route taken by the reply since it must travel along the reverse of the path taken by the request.

Intrinsically, the SSP mechanism implies no particular philosophy in terms of programming methodology; nor is it a reliable means of communication without additional procedures or constraints. A natural way of use is to regard the SSP as a primitive form of remote procedure call (RPC). In many instances the speed of interaction over a local network is such that the extra delay incurred in comparison with a local procedure call is sufficiently small that a straightforward synchronous (blocking) approach is quite adequate. A question raised in the UNIVERSE context is whether this continues to be useful in a network with substantial delay.

In the context of a single LAN, the question of reliability arises mostly in connection with recovery from major component failure affecting one or other of the end hosts, rather than packets being lost by the network. In the UNIVERSE network, however, bridges discarded packets as the only defence against congestion. The primitive RPC mechanism offered by the UNIVERSE SSP was not only unreliable but was limited in the amount of data which could be transferred in either a request or a reply. Generally, to overcome these shortcomings a choice of approaches is possible, such as using multiple packet call and reply messages and either using a reliable network service or implementing reliability mechanisms within the RPC protocol. In UNIVERSE the emphasis was on simplicity. Many RPC-style transactions do not require extensive data exchange, making the SSP quite adequate in this respect for the majority of such interactions; and to achieve reliability, the idea of *idempotency* was exploited at application level.

The notion of *idempotency* is to arrange that (arbitrary) repetition of a given transaction between client and server produces the same effect as a single instance of the transaction. Thus, when a client times out awaiting a reply from a server and has no indication whether it was the request or the reply which was lost, it is nevertheless safe to repeat the request. This contrasts with the two previous approaches by involving the application level in lower-level error recovery. It also requires relatively little software support and was used by several of the distributed systems described in this chapter.

7.2 Monitoring

7.2.1 *Objectives*

Management and control facilities are necessary for the proper administration of any service network. Monitoring facilities are needed to gather the basic information necessary to understand the behaviour of a network and to measure use of network components by subscriber hosts. This information forms the initial basis for diagnosing problems as they occur in operation of the network. It is also fundamental to the imposition of any form of control, accounting, or charging for network use. Finally, analysis of statistics gathered over a period can be helpful in managing a network by showing, for example, where heavy loading is occurring, indicating the need for additional resources or, conversely, extra controls. While general information on loads and traffic patterns is a useful adjunct to fault diagnosis, other types of monitoring, such as error logging from network components, and the use of probes may be more appropriate in relation to faults, both in logging continued correct functioning of specific devices and in tracking down a fault.

The UNIVERSE network offered the possibility of studying the monitoring needs of a multiple LAN network. Since the network encompassed several sites, some opportunity existed to consider the inter-relation of individual site network management in achieving overall network management. Among the activities subsumed under the term network control is that of updating both the network tables and the code for various network processors. In the case of the latter, there are the additional matters of how to develop, load, test, and debug new versions of code for such elements: an activity sometimes referred to as 'life-cycle support'.

The specific objectives of the network monitoring and management experiment included

(a) the gauging of traffic both on an individual LAN and between LANs; in the case of the latter, a further distinction can be made between local LANs (situated at the same site) and remote LANs;

(b) an investigation of probing techniques; and

(c) an assessment by both monitoring and probing of the availability of network components.

To attain these objectives required putting in place a substantial amount of infrastructure devices and code.

The life-cycle support mechanisms used in the project were generally a combination of the facilities provided by the bootserver (Chapter 5) and versions of some of the facilities of the Cambridge system (Needham and Herbert, 1982) exported with the large server (Chapter 6). Their use within the project was not significantly original and was, in practice, usually confined to a single site.

7.2.2 *Structure of monitoring subsystem*

The various entities subjected to monitoring could be conceived of as forming a hierarchy, at the top level of which was the whole network. Next came individual organizations or sites; within sites, constituent LANs were distinguished. The final level of the hierarchy was composed of individual stations or hosts on each ring. In considering the system as a reflection in miniature of an operational system, the monitoring centre (or, more generally, centres) was envisaged to be associated primarily with the whole, that is, sites and their interconnection. However, an additional service might well be the provision of the site level services to or on behalf of those sites with configurations too small to justify their own dedicated monitoring subsystems. While such a service might not be able to provide the same degree of monitoring, it could be realized by rearranging components used in this experiment and would not require any essential redesign.

The notion of more or less autonomous arrangements for site monitoring, with the capability for co-operating with one or more global network analysis centres, encourages the idea of a distributed architecture for this subsystem. This model is also appropriate for the purpose of safeguarding local communications operations when wide area communication is unavailable. In the case of the UNIVERSE network, this aspect was exaggerated because the network architecture imposed fixed routing among sites and the satellite was only available for a short period each day.

Thus, collection of monitoring data and provision of life-cycle support were essentially delegated to site level. At times when connectivity allowed, these local collections could be accessed and transferred to a central site (UCL) for longer term storage and subsequent analysis. These site level collection and temporary storage facilities for monitoring data were provided by the simple file service available on the bootserver (Chapter 5, Section 5.2.5) installed at every site.

The sites subjected to monitoring as part of this experiment were GEC, Cambridge, and UCL. There were three rings at UCL, with a bridge connecting each pair. The network configuration and monitoring resources at UCL are shown in Fig. 7.1.

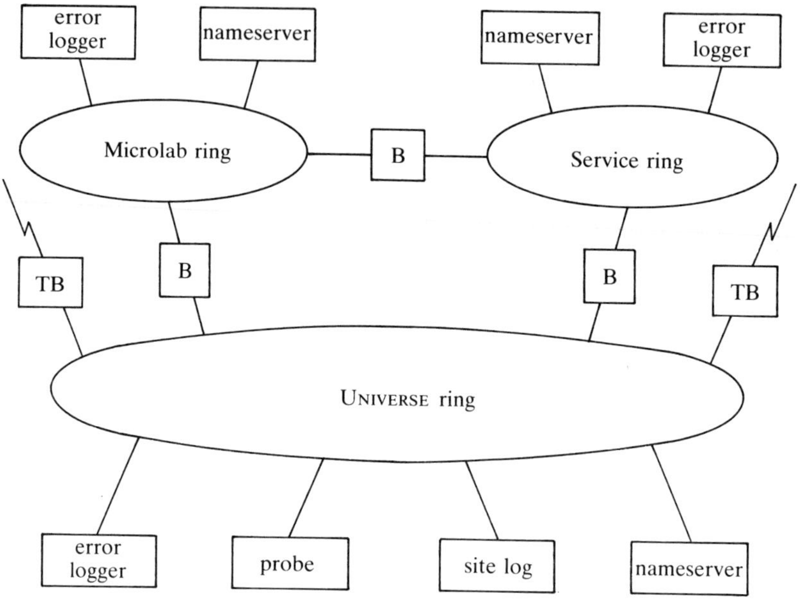

FIG. 7.1 Network monitoring resources at UCL.

7.2.3 *Components*

Even in such a relatively small network as UNIVERSE, there are a substantial number of elements which are candidates for probing and monitoring. This in turn implies the need for a variety of tools and embedded code segments distributed all over the network. A substantial fraction of the effort available to this experiment was absorbed in constructing, coding, testing, and generally establishing the essential components of the monitoring subsystem infrastructure necessary before any monitoring investigations could be carried out. The components of this infrastructure are briefly described here.

(a) *Date and time* An essential part of the process of collecting monitoring data is to stamp items with the date and time as they are collected. Some of the various components of the monitoring system may have their own, individual clocks. For those that do, it is necessary to provide a synchronization clock for occasional reference; for those without clocks, it is necessary to provide a date and time service capable of handling quite frequent requests. In any network it is prudent to provide more than one such reference service, to guard against monitoring components being severed for unnacceptably long periods from a time source. The limited availability of the satellite segment in UNIVERSE served to exaggerate this, as a consequence of which time servers were made available at all sites participating in the monitoring experiment. The existence of more than one time server in a system raises further questions about synchronization. To the degree necessary within the network generally and this experiment in particular, this was ensured by the nature of the time servers provided by the project. These were all based upon the Z80 small server (Chapter 6), to which was attached a radio receiver tuned to the time signals broadcast from near Rugby. Clocks of this sort are quite adequate for the purpose of correlating events recorded at different points in the network. No attempt was made to use these clocks as a means of noting transit delay between sites, nor was there ever any intention of so doing.

(b) *Site log* Access for writing to the monitoring data storage facility provided on the bootserver at each site was usually by use of the site log service, to which an SSP request containing data to be logged could be sent. The received data was stamped with the date and time and filed in the current day's log; depending upon its nature and urgency, it might also be displayed on the system console screen. Read access to the log, to transfer data from several sites to a central system for collation and further processing, was provided by using the general fileserver access primitives (Chapter 5). This logging service had much wider applicability than just the collection of monitoring statistics. Any system having routine events to log in the course of operation could make use of it. A number of the experimental components of the network needed directly attached consoles not only for diagnostics but also for displaying regular operational information. It was observed that this led rapidly to the need for a large number of consoles and space to house them; this was particularly noticeable at UCL which had a more complex network topology. In any more operational network, greater use of the site log by network components would become mandatory.

(c) *Satellite status* An alternative approach to providing access and display of the status information on the satellite segment of the network was implemented towards the end of the project using the MONITOR-BROADCAST channel provided by the satellite bridge. A supervising service monitored the status of the network and broadcast the information regularly, sending an item every 10 s. Satellite schedules and messages were also broadcast. A program was written for the BBC microcomputer which could be loaded when required to display each item of information pictorially as it was received. Because of the multicast reception facility, one or more BBC microcomputers at any site could run this program simultaneously. The attraction of this scheme for delivering operational information is that it is available to any display station situated anywhere on the network. A point of technical interest was that this was the only system which used the broadcast service with multiple transmitters.

(d) *Error log* As well as a site log service, a Z80 (small server) error logging service was provided at those sites participating in monitoring. The functions implemented in the error logger included

- emitting regular probe minipackets to every station on its ring and constructing a bit-map status record of the stations observed (from the response bits) to be present and powered on; whether the ring was on or not could also be detected;
- collecting fault minipackets from nodes, some interfaces (e.g., Polynet VMI-1), and the monitor station; and
- filtering repeated minipacket reports, especially the stream of ring reframe notifications associated with a ring restart and also the fault reports from a station which result from a cable break.

All error logger records were date and time stamped, using an internal clock regularly synchronized with the local Rugby clock service. Under control from its 'owning' monitoring centre, the error logger normally reported all the data it gathered to the local site log service, though it could be directed elsewhere (e.g., to the fileserver at Cambridge).

The UNIVERSE error logger was developed at UCL and was based on a Z80. Special hardware and software was used to provide the capability, in a single station, to function normally on a chosen non-zero address, as well as to receive fault minipackets (which are always sent to station zero). (*Note:* The need for these special arrangements arises from the special meaning attached to setting the station select register to zero; namely, not receiving. Thus, minipackets from station zero can only be received by stations selecting 255 (receive from anywhere). Since all known implementations of the basic block protocol only receive from one station at a time and use the select register to ensure this, station zero is effectively unable to transmit basic blocks to anywhere.)

(e) *Probes and reflectors* A traditional tool for network verification and diagnosis is a set of strategically placed 'probes' and 'reflectors'. By 'bouncing' probe packets off various reflectors, the state of connectivity or, alternatively, the points of failure of the network may be established. No special-purpose reflectors were built into the network but all bridges and nameservers did have the so-called *code-id* service (Chapter 5), which would respond to an SSP request by sending an SSP reply to the originator containing the name and version of the code being executed. Using the *code-id* service as a reflector, it was generally possible, by probing successive bridges and nameservers, to establish the state of specific paths in the network.

(f) *Host and bridge statistics* Another common mechanism for statistics gathering is to insert code segments into the component to be monitored. This was done for all the bridges and for one of the hosts, the GEC 4065 at GEC. The principle was the same in all cases. The code was controlled from an external device (by SSP commands) and some of the parameter values could be varied: gathering could be enabled or disabled, rates associated with both gathering and divulging data could be varied, etc.; and the data could be retrieved. Both polling and regular unsolicited datagram transmissions were used by the bridges to report such data.

7.2.4 *Results*

Gathering statistics and probing the network became a regular activity during the latter part of 1983 and was continued into 1984 locally at UCL. A reasonable body of data was collected over this period, although there are gaps, arising generally from temporary failures of essential components such as a satellite bridge or a bootserver and its associated site log. Here, we attempt to summarize the salient features of the results; for complete details, the reader is referred to the project report on the subject (Report 3; see list in Appendix B).

Use of the local networks at Cambridge and UCL was not restricted to project participants: both had significant use by student user populations. The level of activity resulted, for example, in about 2000 site log events per day at UCL.

(a) *Bridge traffic* The sizes of blocks passing through bridges were monitored. In the ring–ring bridges, the bimodal distribution reported by Shoch and Hupp (1980) was not present in all cases, though there was a predominance of short blocks. Two of the bridges (one each at UCL and Cambridge) were carrying service traffic; the bimodal distribution was present at UCL but not at Cambridge. This difference may well be related to the different distribution of components and styles of use at the two institutions. At Cambridge, there is a substantial amount of file access traffic passing through the bridge; at UCL, service traffic through the bridge was predominantly terminal access, via concentrators connected to one ring, to systems connected to another ring. A significant amount of traffic over the satellite segment was carried in large blocks, corresponding to the predominance of various forms of experimental and routine bulk data transfer. Also monitored in the bridges were the types of traffic: SSP, datagram, or LVC. A significant difference between UCL and Cambridge bridge use was the preponderance of SSPs at Cambridge, compared to OPENs at UCL. This is in keeping with the different styles of use.

The most significant observation from the point of view of the network architecture was that the measured rate at which the ring–ring bridges dropped blocks was small enough to lend support to the UNIVERSE protocol assumptions. The rate of loss through the most heavily used bridge (at Cambridge) was around one block in 3000 or about 0.03 per cent and was quite steady. The peak value of the hourly load on this bridge was about 90 000 blocks or 25 block/s. These observations can be turned around to mean that the performance of the UNIVERSE ring–ring bridge was adequately matched to the demands made of it by the number and type of clients using it.

(b) *Host statistics* Using the host statistics collection system implemented by GEC, measurements were made on basic block, SSP, and BSP activities. A notable feature of the results was that approximately half of all the SSP traffic was to the nameserver. Moreover, a single host was recorded generating 10 SSPs per second to the nameserver. The architecture is specified to allow the use of cacheing for the results of name-lookup requests; of course, in simple experimental situations this was frequently omitted. The results point up the importance of using cacheing in any more service-oriented implementation.

(c) *Database server* Use was made of a commercial database package (MISTRESS) to assist in organizing the monitoring data. Collection throughout the monitoring system was a two-stage process: first into a fileserver and then, under more manual control and in larger segments, into the database. This was adequate for use in the project but would not suffice for regular, routine operation in a service network. Undoubtedly, it would have been a great advantage to have used a 'database server' for this purpose, to combine the two steps into one, more automated operation. However, the requirements of such a system are severe. To be able to cope with the expected level of activity, it needs to be fast and to have record-level locking. This is necessary to support simultaneously the rapid transaction rate generated by record update and addition during data collection together with the provision of data selection and retrieval, on demand, for analysis, fault diagnosis, etc. To provide essential selection and manipulation support, facilities for date and time stamp comparison and conversion to and from duration (relative to a specified date and time) were found to be particularly important.

(d) *Probing strategy* This worked quite well because the path taken by the probe can be predicted from the fixed routing used in the network. The mechanism is not a satisfactory one when there are loops in the network topology, as at UCL. The fixed routing meant that one of the three bridges connecting the three UCL rings in a loop could never be traversed without introducing another probe on one of the other rings. This requires the data from the two probes to be correlated.

Another possibility considered was to introduce a set of servers at strategic points in the network which would onward route packets containing complete source routing information. This was never done; it is somewhat analogous to the 'chain-letter' mechanism, described in Chapter 4, used to provide loading traffic for the satellite segment.

To provide a regular and complete view of the state of connectivity of the network, the two or three probes used in the experiment would not be sufficient, since they were liable to be severed from parts of the network. To cope with a larger network or one with dynamic routing, a more general and automated way of linking the probing with the network topology and routing tables would also be needed. Use of source-routed probe packets would also seem to be necessary. All these considerations tend to suggest that for this method of maintaining a continuous picture of the state of network connectivity to be effective requires that the probes and reflectors be amalgamated as integral parts of the network components.

7.3 Use of remote fileserver

A topic of continuing interest in the development of distributed computing is the provision of file storage. Generally, the idea is to have a machine with attached disks, whose sole purpose is to provide safe storage of data for the benefit of other systems on the network. Physical storage space management, the making of regular copies for backup, and provision of access to the data become the dedicated roles of this system. Particular implementations of this concept vary greatly in the details of what facilities are provided. All are intrinsically multi-access servers; some provide secure access; the structure and style of accompanying directory mechanisms varies widely: some provide low level, primitive storage of sequenced bytes; others provide end-user, tree-structured directory systems.

The fileserver used in the Cambridge distributed computing system is of the low-level sequenced byte storage variety (Dion, 1980 and 1981). Special attention has been given to providing both security and optional facilities for transaction processing with well-defined commitment points, that is, atomic transactions. If needed, clients of the system impose their own directory and record structures on this storage in a manner independent of the fileserver. A substantial and important set of fileserver clients consist of the set of processor bank machines of the distributed computing system (Chapter 6). (A full description is given in Needham and Herbert, 1982.)

The machines of the processor bank generally run the TRIPOS operating system (Richards *et al.*, 1979). In early versions of the system, each client machine contained the complete code to implement the TRIPOS file structure above the primitives provided by the fileserver. Each client operated a cache of file blocks but the need for continual access to a shared directory system undergoing frequent updating imposed a substantial overhead of small transactions on the fileserver. The latter had originally been designed to give high-speed access to more substantial amounts of data. Under the load of more frequent, smaller transactions, it rapidly became a bottleneck. This situation led to the development of the TRIPOS filing machine (Richardson and Needham, 1983).

Apart from the problem of handling a large number of requests for small amounts of data, there was another aspect of the situation which placed both the clients and the fileserver at a grave disadvantage. The shared client filing system requires transitory but exclusive locks on parts of the directory structure while it is updated during client file manipulation and access. This requirement for exclusive access reduces the effectiveness of any scheme for individual clients to cache parts of the directory structure; this structure is in continual use for interpretation of file names and the need to keep refreshing each client's cache leads to substantial load and performance degradation. What is even more unfortunate is that much of the traffic is in practice nugatory since that part of the directory in use at one client has probably not been altered by another; this is a result of locks only being available on inappropriately large sections of the structure, each section being the content of a file in the fileserver.

The source of this problem lies in an attempt by a number of clients to cache individually what is in fact a genuinely shared data structure, the directory system. The approach taken to alleviate this was to provide a dedicated system for managing this shared data structure: this system was the filing machine. Only one filing machine per fileserver is (properly) allowed in the current architecture and this manages the directory-level locks, so that the same problem does not arise in the filing machine (in contrast to the situation for individual clients) cacheing substantial amounts of the shared directory structure. In fact the filing machine also managed access to individual files, performing both read-ahead, asynchronous write, and substantial cacheing on behalf of its clients. The general arrangement for clients accessing files kept on the fileserver via the filing machine is shown in Fig. 7.2.

Qualitatively, the implementation was successful in that it was able to support up to some 15 processor bank systems before performance degraded unacceptably. This was a substantial improvement over the previous situation. The rate of data cache read requests was in the region of 5000–25 000 per hour; the cache hit rate was typically about 90 per cent.

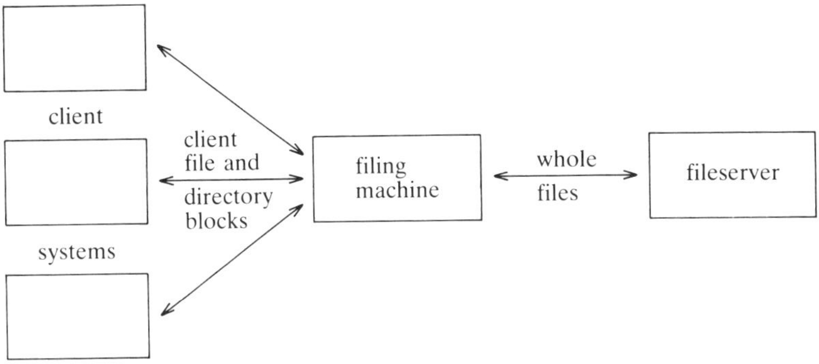

FIG. 7.2 Local use of filing machine.

7.3.1 *Remote access*

An analogy which has been used in describing the operation of the filing machine is to remark that it provides a 'retail' service to its clients, while itself making use of the 'wholesale' service offered by the fileserver. In the context of the UNIVERSE network, the existence of the cache in the filing machine, coupled with its making relatively infrequent but substantial 'wholesale' requests of the fileserver, all suggested that the arrangement might continue to function acceptably, even with the satellite segment interposed between the filing machine and the fileserver. Although the satellite delay would probably be highly detrimental to the small transactions emanating directly from a client, the satellite channel had high bandwidth and the effect of cacheing might enable the filing machine to mitigate for the user the worst effects of delay. This then was what was to be tested.

The general arrangement is shown in Fig. 7.3. The experiment was made possible by the availability at RAL of TRIPOS systems based on Intel 8086 Multibus hardware. One of these systems was configured to be able to boot itself with a system similar to that used in the Cambridge processor bank, so that it used the filing machine instead of its own disk. A 68000 system was provided at RAL to act as a filing machine, accessing the fileserver at Cambridge for its files. Later developments enabled a client system to make a dynamic connection to the filing machine, in addition to using its own disk. This allowed simultaneous access to both local and remote filestores.

As an aside, it is interesting to note that because the development environment for this experiment existed at Cambridge, it was decided to boot the filing machine from Cambridge, using the satellite. This was arranged by making temporary alterations in the nameserver's tables at

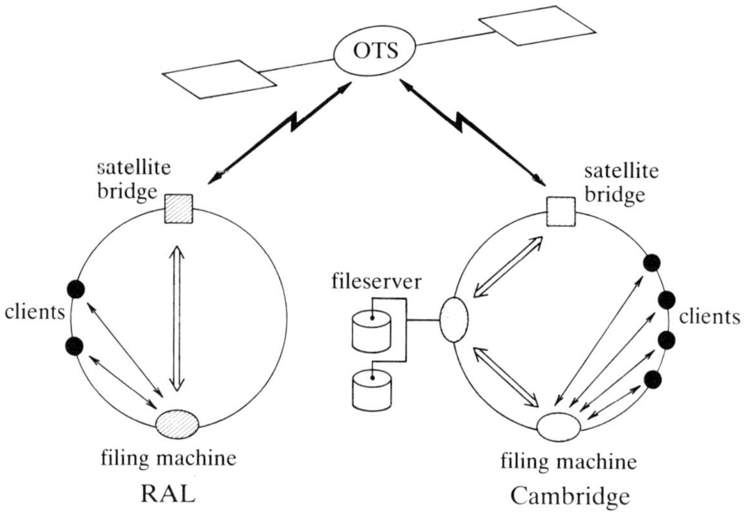

FIG. 7.3 **Remote access to fileserver using filing machine.**

RAL to point to boot services at Cambridge. Though not without its problems, it was nevertheless a tribute to the ability of the network as a whole to behave as an extended LAN that this was not only possible but was in many respects to be preferred (for the purpose of the experiment) to providing otherwise-essential environment at RAL.

7.3.2 *Experience and conclusion*

Initially, only a single system at RAL made use of the filing machine. Later, both large servers (Chapter 6) at RAL were able to mount this filing system, in addition to their own local, disk-based file systems. With only three machines using it, the filing machine was very lightly loaded, both with respect to processor capacity and amount of cache activity. The most notable effect of this was that, in the case of the disk-less system at RAL, the code for most of the TRIPOS commands became permanently resident in the filing machine cache. (The large server systems used this system in such a way that command code was usually loaded from the local disk.)

Experience of this mechanism was much as might be expected. Simple operations or those which required synchronization with the fileserver at Cambridge were relatively slow. More complex operations, especially those allowing scope for use of the cache, would typically take longer to begin but would thereafter proceed at rates comparable with those obtained for use of a local fileserver. Thus compilation or editing, once started, were not noticeably slower, whereas directory examinations were considerably delayed since they typically involved not only path-name resolution but access to the header block of each file.

Richardson and Needham (1983) give details of some timing measurements made of a variety of accesses to the filing machine. Briefly, the access times for requests made by the filing machine of the fileserver took between 2 and 7 times as long over the satellite. For operations requested by a client of the filing machine the variation is greater, from little different (for read and write) to a factor of 25 (for delete).

One of the authors (CSC) made quite substantial use of this mechanism in aid of maintenance of the large server systems, most of whose code originated in Cambridge. Subjectively, the convenience of the functional completeness offered by this mode of accessing the remote filestore far outweighed any irritation caused by the slowness of some of the simple commands.

For the purpose of the experiment, the system was arranged only for read access to the Cambridge filestore by machines at RAL: no mechanisms were developed for sharing locks between the filing machines at Cambridge and RAL, nor for exchanging information about invalid cache entries. The experiment was generally judged a great success, sufficiently so that some thought was given to remedies for these shortcomings. For several sites to share each other's filestores, the simplest practical configuration would be for every site to have both a fileserver and a filing machine. Each filing machine would naturally access its own fileserver; it would also require some form of shared access to the fileservers on other sites, for which solutions to the lock and cache synchronization problems need to be found. Improvements would also need to be made to the performance of simple commands in any system for more general use. More generally, in order to be able to realize the potential of distributed systems to exploit replication to increase capacity or improve reliability, it would be necessary to enable a number of filing machines to share access to a number of fileservers. No implementation work was carried out in the project on any of these aspects.

7.4 File distribution

Basic to the use of networks in computer communication is the transport of bulk data. Increasingly, it is important to be able to transfer a wide variety of digital data: examples range through voicegrams, bit-map images (including facsimile), graphic displays, teletex, videotex, and the more conventional material such as text and computer programs. For communication between widely dispersed systems of personal workstations, it may be important to shield the user from details of the transport mechanism, temporary severance of parts of the network, and temporary unavailability of the remote station. The cause of the latter may be component failure, dedicated use of the personal computer, time-zone difference, or different work patterns. For all these reasons, it is desirable to be able to delegate the task of transfer in the confidence that, even if it may be delayed, responsibility will be taken for delivery and the user may proceed in many respects as if it were already accomplished.

7.4.1 *Background to experiment*

The file distribution system was conceived not only as a system of intrinsic potential interest but also as a test of the ability of the network to support distributed systems and to provide a suitable vehicle for the development of bulk transfer mechanisms appropriate to the satellite segment.

Several important properties of the UNIVERSE network stemmed from the use of a broadcast satellite to achieve wide area coverage. Of particular relevance were

(a) the possibility of direct use of broadcast to enable simultaneous transmission of data to several sites;

(b) the relatively high cost of the satellite segment, making it desirable to optimize its use; and

(c) the incidental property of the network being severed more often than not because, being an experiment, the satellite was available on a very limited schedule.

In a service situation, a requirement to use off-peak, cheaper rates could create a need for a system with operational characteristics similar to those necessary to cope with property (c). Optimized use of satellite bandwidth and the broadcast facility would both be important in providing a useful and economic network service. The experiment focused on two aspects: the basic architecture of a distributed set of co-operating systems, called *agents*, whose purpose was to provide an end-to-end file transport service capable of withstanding the intermittent availability of network connections; and the study of novel techniques for data transfer over the satellite segment. This section describes the first of these activities. The experiments in data transfer over the satellite are described in Chapter 9.

The file distribution system was designed with several candidate applications in mind. Within the project there was a need to distribute system software in the form of loadable images to the bootservers at every site. Provision of printing services of various degrees of sophistication could also be based on such a system. As originally conceived and implemented, the file distribution system was in principle not concerned with the content of files and could equally well form the basis for facsimile transfer as for computer or wordprocessor generated document transfer. A particularly appealing direction envisaged for development was that of mixed-media document transfer, where documents might consist of a combination of computer or facsimile text, bit-map or graphic pictures, and voice commentary. To this end, the service was designed to convey bland sequences of bytes: it was to remain ignorant of any structure or information which might be embedded in the data.

It was an assumption that individual LANs were very reliable and that the local representative, or *agent*, of the service would always be accessible to its clients. The service as a whole was to be robust against both the severance of communications between sites and intermittent availability of client hosts. One strategy available to assist in providing reliability is to arrange for intermediate copies to be made of the data under transfer. Since holding these copies within the file distribution system also provided a natural way of separating local and long-haul transfer protocols (so that the latter could easily be the subject of experiment), it was decided to limit the investigation to just such a file-level store and forward system.

7.4.2 *Overview of architecture*

The design of the file distribution system may be described by analogy with the conventional shipping agent. The client engages the services of such an agent who undertakes to transport goods between specified locations. In doing so, the client's agent may make use of one or more peer agents (typically in other countries) to deal with remote ends of the operation, such as collection or delivery. Transfer between countries happens between agents, typically with the aid of an international carrier.

The experimental file distribution system was based closely on this idea. A set of agent systems with associated private disk storage were available on the network, situated so that there was at least one at each site. Client hosts making use of the system were registered with a particular agent which acted for them when they were involved in transfers, whether as originators or recipients of data. A client system wishing to send one of its files to another system contacted its local agent, instructed it about the destination and co-operated in transferring the data to the agent. The responsibility for delivery thereafter became the agent's. The agent would in general contact the recipient's agent and then, using appropriate satellite segment transport mechanisms, the data was transferred to the recipient agent. The latter would in due course deliver the file to its client and then notify the originating agent, which in turn informed the original client of the successful transfer. A similar arrangement was used to enable a client to fetch a file from a remote system. In fact the services offered by the agent system were slightly more general: an initiating client could cause a file to be transferred between two other host systems by making a request of its

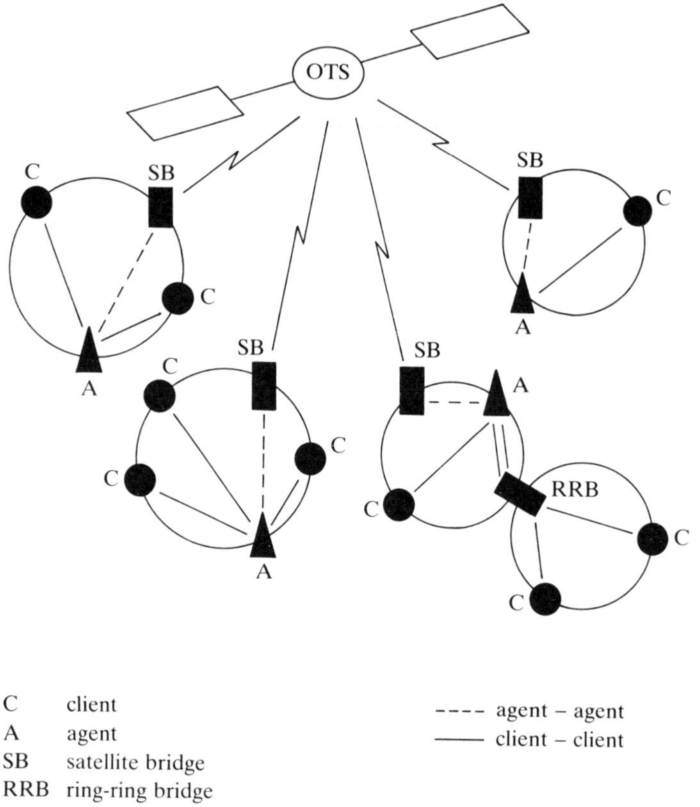

FIG. 7.4 General arrangement of clients and agents.

own agent. The agent contacted the agent of one of the hosts involved and the transfer then proceeded much as before. The report on the outcome of the end-to-end transfer between hosts was sent to the client which made the original request. The general arrangement of agents on the network is shown in Fig. 7.4.

To be satisfactory, such a system needs to be both reliable and available. To achieve the latter requires

(a) agents to be placed sufficiently 'close' to their clients as not to be cut off by intermittent network availability;

(b) sufficient capacity to avoid unavailability due to congestion;

(c) reliable agents; and

(d) reliable end-to-end transport of files.

In the experimental system, a single agent was placed at each site to satisfy (a) since the local networks were sufficiently reliable and the satellite segment was scheduled to be available only occasionally. Achievement of (c) relies essentially upon good software and hardware engineering, assisted in this case by use of dedicated systems for the agents. Capacity was not studied in this experiment.

The aspect of reliability which received most attention in the system design was item (d), robust communication of files, both between the client host and its agent, and between agents. Typically, file transfer protocols (FTPs) involve the exchange of two distinct forms of information: protocol control messages and bulk data. Procedures to ensure that parts of the latter are not lost or disordered are well known; they generally involve sequencing and retransmission and are often incorporated into one or more of the classes of service offered by the network. Also included (though not in the network service) is some method of signalling the end of the data, so that the recipient knows whether it has successfully received everything. A variety of protocols which adequately provide these facilities for bulk transfer were available to the agents.

(1) *Command and control interactions* An important function of the control messages of a conventional FTP (such as that defined in DCPU (1981*a*), for example) is to prevent total loss or duplication of a file during transfer. In the agent system design, this aspect of control was removed from whatever FTP was in use for each stage of the transport and was instead carried in separate transactions, which established the identity of the transfer at the start and which party had responsibility for the file at the end of the transfer, depending on its status.

These transactions became part of the set of system control interactions which were all implemented using idempotent SSPs, as described in Section 7.1. An example of the technique occurs when a client host establishes a new commission with its agent. Commissions were identified within the issuing agent by the assignment of a sequence number. They were identified globally among the set of all agent commissions by being tagged additionally with the issuing agent's identity. The host, of course, required both acknowledgement of the commission and knowledge of its identity. The initial SSP to an agent contained all the details of the desired transfer; the reply contained the identity. However, at this stage the commission was held and would not be executed by the agent. A further SSP was required from the client, quoting the commission identity, to release it. If the reply to the first of these was lost, the commission could never be released and the agent simply timed it out and deleted it; the client, of course, also timed out and repeated the request, establishing a commission with a different identity. If the reply to the releasing SSP was lost,

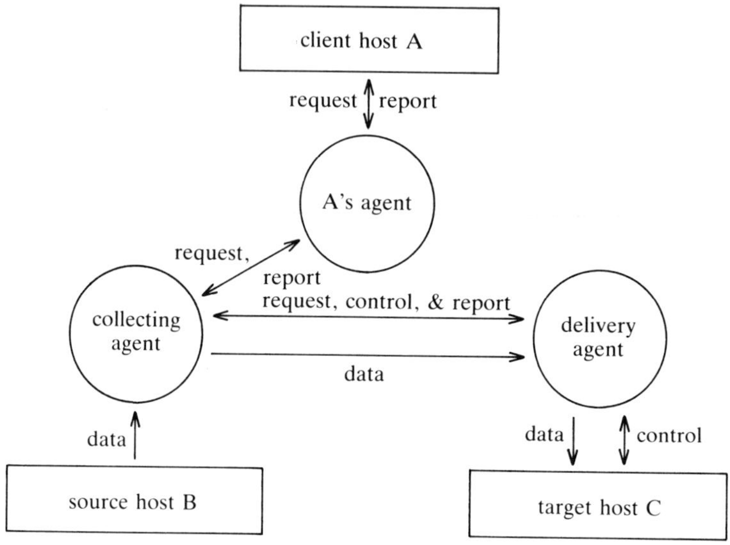

FIG. 7.5 Third-party transfer. (© *1984 ACM*)

it would be repeated until a reply was received; since releasing a released commission was evidently harmless, this SSP was trivially idempotent.

Other control functions between clients and agents included the final report, status enquiry, logging and diagnostic control, and agent startup and closedown.

The most general set of interactions among agents arose in the case of 'third-party' transfer, which is shown in Fig. 7.5. Client host *A* requests its agent to arrange the transfer of a file between hosts *B* and *C* and the agents for *A*, *B*, and *C* are all distinct. Agents use idempotent SSPs to pass the commission to another agent for the next stage of processing: typically, *A*'s agent passes the request to the collecting agent. Once collection is complete, the collecting agent passes the commission to the delivery agent. However, the collecting agent does not drop out of the chain. The final report on the outcome of the transfer passes back along the chain (using idempotent SSPs as usual), triggering deletion of the commission records in each agent. The forward links in the chain are used to obtain the latest status of a commission.

(2) *Collection, transfer, and delivery* For all bulk transfers between hosts and agents or between agents, the initiative was given to the receiving party. This simplified the control logic in the agents which made use of a variety of procedures for bulk transfer, in which some of the experimental

forms could only be initiated by the recipient and had no internal mechanism for communicating to the sender the final status. For collection, the activity was initiated by the collecting agent. The action was one of copying the file: there was no requirement by the agent for the source to delete the file or perform any other action after the copy has been taken and so the agent need not inform the source when it has read the file. An application requiring such action upon successful delivery would make arrangements outside the service provided by the agent. This represented the simplest option generally required of transfer systems. If deemed necessary, additional control messages could be added to the system without any fundamental change to the architecture. Thus, for collection, no control messages were required. The agent simply used whatever procedure was defined for collection from the host service to obtain its copy of the file. The procedure might incorporate recovery mechanisms but these would be transparent to the agent. If the procedure reported failure of a temporary nature, the agent would retry from the beginning.

For transfer between agents, two experimental procedures were developed, point-to-point and broadcast. Both of these were designed explicitly for transfers between systems with random access storage. They also took into account the comparatively long round-trip delay introduced by the satellite segment. The broadcast transfer made use of the physical broadcast properties of the satellite. The independence of the agent-to-agent transfer stage from the hosts made it easy to experiment with various mechanisms for this stage. It also, incidentally, allowed the agent implementation to proceed in parallel with these developments, which are described in Chapter 9. Once the receiving agent had satisfactorily received the file, it sent a control message (via SSP) to the sending agent, which caused the latter to delete its copy of the file (but not the commission record).

The normal sequence of events during delivery began with the agent notifying the destination host that a file was available. The host then transferred a copy from the agent to itself and notified the agent upon completion. The notification by the host that it now had a copy of the file caused the agent to delete the file from its own storage and to begin the process of notifying the requesting client of the successful outcome. In the UNIVERSE environment there were several desirable destination services (notably print servers) using procedures with the general properties that they had to be initiated by the sender and they incorporated satisfactory methods of notifying the sender of the outcome of the transfer. Knowledge of how to use this class of procedures was also easily incorporated and opened up a range of services available via the agent without the need to alter those services at all.

7.4.3 *Implementation*

The agent was implemented on a large server system. At least one such system was available at every site. All programming for the agent system was in BCPL, the language integral to TRIPOS.

The agent consisted of a number of co-operating tasks, which communicated by exchanging message packets. A spool task provided controlled access to a set of spool entries, each representing an agent commission. When action was required for a commission, it was passed to an executive task, where the main agent logic was implemented. This task was responsible for generating control SSPs to hosts and other agents. It also managed the loading and execution of bulk transfer code implementing collection from and delivery to hosts, and transfer between agents. (The actual transfer processes executed as separately scheduled tasks of the operating system.) The responding end of any transfer was handled by a task which was permanently listening and managed the loading and execution of code according to information in the OPEN. The structure was completed by a (small) number of identical tasks which handled incoming SSPs from hosts and other agents. As well as handling control messages, these tasks implemented remote diagnostic and debugging aids for the agent.

Two files provided the information needed by the agent system to enable it to decide which agent to contact for collection or delivery of a file. The set of agents co-operating to provide the service was defined in a file called the agent configuration file. Each entry in the file bound the UNIVERSE global name of an agent system to a unique integer, its identity. This identity had two uses: as a constituent of a commission identifier and as the permanent identifier associated with the contents of the particular agent's spool storage. The latter was established at cold start, when the spool was created, and was checked upon the occasions of subsequent warm starts. The configuration file was the only place where the identity of an agent was bound to its network name. Adding or removing an agent from the system was essentially a matter of inserting or deleting the corresponding entry from this file and then transferring the file to all other agents, so that the replicas were updated. While adding an agent could be done at any time, in the prototype system, agents could only be removed when the system as a whole was quiescent and the agent in question was empty. The indirection provided by the configuration file also allowed for the possibility of moving a particular agent to a different machine in case of, for example, machine failure.

A second file, the host directory, served several functions, the primary one being to register with an agent every host or service participating in the file distribution system. Strictly, a name registered in the directory was a partial or complete UNIVERSE network name of a service available on a participating host. The names were public, in that a client host used them to specify the acquisition and delivery services to be used at the endpoints of a file transfer. The agents used the directory to determine which agent should handle the acquisition or delivery of a file. Also recorded against each name were the collection and delivery procedures to be used. In some cases, one of these procedures was null because the client system only supported either collection or delivery. A print server is such an example. Addition or removal of a host service from participation in the agent system was essentially a matter of inserting or deleting the relevant entry in the host directory.

Both the agent configuration file and the host directory were global, that is, every agent needed access to the same file and the form of the entries was independent of situation on the network (e.g., all names were fully qualified and all agents used identical sets of procedures). Every agent obviously required access and the simplest way of achieving this was to replicate both files in full in every agent.

The system design was elaborated during the summer of 1983. The main implementation and testing effort occurred during November and December 1983, after the World Telecom'83 exhibition in Geneva. Attempts were made during December to install the system at all seven sites. This coincided with a period of substantial unavailability of communication with the BTRL and UCL sites. The latter also affected communication with Logica. Installation was never completed at these three sites. All testing and development were accomplished using the set of 16 systems and services listed in Table 7.1 and shown in Fig. 7.6.

The large servers implementing the agents at each site could also act simultaneously as clients of the file distribution system. A command interface was written for TRIPOS and all transfers and control operations to agents were initiated by commands from 8086 or 68000 TRIPOS systems at RAL.

Apart from commands for file transfer between any two file distribution clients, simple commands of the debugging and management variety were available to start and stop agents, to switch logging files so that the previous one could be transferred and printed, and to control debugging messages sent to the agent system console.

Each agent was registered as a client of itself (in which case file collection amounted to a copy operation and file delivery a rename). This allowed the agent system to act as the conveyance mechanism for updates of its own system files. While this obviously has its dangers, it was effective in an experimental environment of limited duration.

Several of the systems used for testing only had server-oriented software for file transfer. Notable among these were the BBC micro and boot-

TABLE 7.1 Systems used in test configuration

Site	*System or service*
Cambridge	Bootserver
	Laser printer
	BBC micro
	Large server client and agent
GEC	Bootserver
	Large server client and agent
LUT	BBC micro
	Bootserver
	Large server client and agent
RAL	BBC micro
	8086 TRIPOS system
	Bootserver
	Large server client
	GEC 4090 as client
	Print service on GEC 4090
	Large server client and agent

server systems, as well as the print servers at Cambridge and RAL. Transfers among systems like these obviously require an intermediary, which the file distribution system provided.

While building and testing of the system was not completed beyond the stage of what might be termed an 'experimental prototype', nevertheless some significant tests of both functionality and robustness were achieved. Transfers between all 16 of the systems listed were not attempted (there being 120 distinct such operations, ignoring direction of transfer); however, sample transfers covering most of the significantly distinguishable cases were made. Simple checks of the system's functions were made, particularly of its performance when hosts or parts of the network were temporarily unavailable. The system acquitted itself satisfactorily in such specifically arranged tests: by the end of the project, the system was ready for preliminary testing by users.

7.4.4 *Evaluation*

In the experimental system, several important features, essential in any service product, were omitted. The whole area of authorization, authentication, security, and access control were not addressed in the prototype system described here. However, experiments in methods of providing two

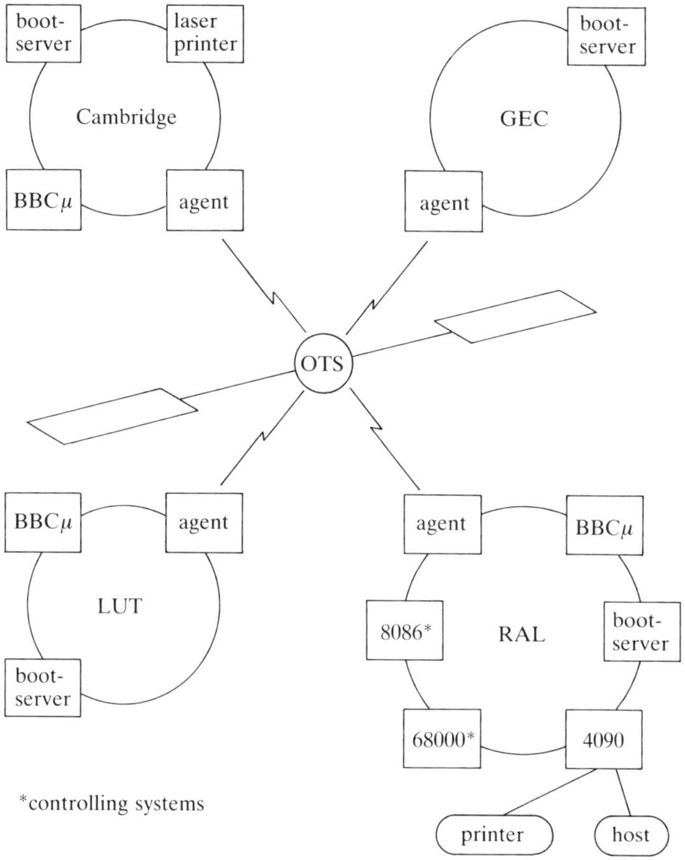

FIG. 7.6 **Test configuration of clients and agents.**

of these elements of functionality, applicable to such a distributed system as the agents, are described in the next section (authentication) and Chapter 9 (encryption). Methods for the management of authorization and authentication have been studied elsewhere, notably in Grapevine (Birrell *et al.*, 1982; Schroeder *et al.*, 1984), and also in Girling (1983). One of the more immediate uses for these facilities is in charging users of the service. In this regard, another aspect omitted from the model was any concept of quality of service. For example, it might be appropriate to offer some level of option on the trade-off between rapid, expensive delivery and slower, cheaper transfer; overnight delivery by any means would be another variant.

A production system would need to provide a much more satisfactory means of managing the registration and configuration databases. Both of these were replicated in every agent and, while this ensured their availability, it also raises questions about maintenance and consistency. One interesting possibility would be to make use of network bandwidth to distribute complete, new versions of the database, rather than a sequence of updates. A related question requiring examination is the tolerance of the system to copies of the databases being temporarily out of step. Exemplary mechanisms for dealing with such matters were developed for Grapevine and were not replicated as part of this experiment.

An issue of product engineering is that of scaling. Individual agents in the system were constructed on the basis of a maximum file size of some 1–2 Mbytes and a maximum of 100 spool entries available on each agent for recording commissions (though these limits could not both be met simultaneously). Neither of these limits is fundamental in the context of this experiment, although the movement of very big files could lead to the requirement to avoid spooling altogether. A sufficiently large amount of activity, generated by a combination of a large number of clients, a rapid commission transaction rate, and large files of data, leading to many files being in transit simultaneously, would require a file distribution system of larger capacity. This could notionally be provided by either having larger agents or, more naturally, by increasing the number of agents. While neither would be a fundamental change to the system architecture, the issues relating to scaling the numbers of agents were not studied in this experiment.

Generally, considerable attention was accorded aspects of the architecture affecting overall reliability. However, one such aspect, the case of agent failure, could be improved. The most obvious and important remedy to the architecture would be to allow a client to use more than one agent to undertake transfers. This was done, for example, in Grapevine. This complicates the registration mechanism and also demands a more general way of locating resources, neither of which were the immediate purposes of the experiment, hence their omission.

A promising sign of the robustness of the system design occurred when both the nameserver and the satellite bridge went down for about 10 minutes at Loughborough during a transfer between a BBC micro at Loughborough and a TRIPOS system at RAL. The Loughborough agent was also restarted during the general process of recovery. This happened during testing of the agent system by one of the authors (CSC) and by the time it was realized that the delay in completing the transfer was caused by the temporary collapse of the Loughborough segment of the network, rather than some agent fault, the file had already been delivered correctly. While the system was intended to withstand network unavailability, it was interesting to learn that it could under some (though not all nor indeed even most) circumstances also cope with loss of the spool containing commission records in one of the agents, when more than one agent was involved in the operation. The ability to do this was not designed into the system: it would be interesting to examine whether adding such a measure of intelligent reliability would be a small design alteration or would require a major rethink.

The potential for reliability exhibited by the system, even in prototype implementation, owed much to the separation of control from data and to the use of SSP interactions in conveying that control. The idempotent SSP is an example of a technique which not only took advantage of the properties of the UNIVERSE network but was also simple to use. This arises partly from the nature of its use being sequential and partly from the property that the SSP request and reply are each indivisible communication units. The experience gained here certainly recommends the use of this style of interaction for passing control in a distributed system.

The activities of the file distribution system had some features in common with a distributed electronic mail system, and it is to be expected that an implementation of the latter could be based on the former. Recent standards work for electronic mail and messaging applications, notably the X.400 series, have introduced the abstract concept of the *user agent* (CCITT, 1987). The agent described here may be viewed as a primitive realization of a distributed form of this abstract concept. Another possible area for application development would be to distributed forms of job transfer system based on the ideas described in DCPU (1981*b*) and ISO (1985*a* and *b*). Such a development could form the basis for a batch segment of a distributed computing system.

An issue which would arise in the development of such higher-level functionality would be the interpretation of a file being transferred. It was a principle of the system that the agents did not need to be concerned with this: a file was merely an ordered set of bits. Ultimately, complete interpretation is an end-to-end matter; however, it remains an open question whether conveying the information to enable interpretation should be entirely outside the scope of the system or be carried in a part of the commission specification only partly interpreted by the agent. It is generally the case that, for the purposes of implementing higher-level services, it is very useful if the control structures of lower levels contain fields reserved for carrying higher-level controls. Constructing a higher-level or more application-oriented agent then becomes equivalent to the agent being able to act on the basis of additional information in the control structure and possibly also being able to perform some processing of the data. (A simple example would be character code translation: a matter which in the original design was entirely confined to the client hosts.)

The prototype file distribution service developed was a store-and-forward one at the granularity of a file. Voicegrams and bit-map images are typically much larger than text documents or programs. An interesting possibility, suggested by the need to handle such large files, would be to extend the architecture to the management of file transfer 'by reference', rather than 'by value' as at present. Here, 'by value' refers to the store-and-forward mode of operation actually implemented, in which a copy of the file in question is taken by the agent system. A system which operated 'by reference' would instead concentrate on alerting the two end-systems at a time when both were available for bulk transfer, thereby suggesting an opportune moment for the transfer to take place; the actual transfer would be directly between the end-user systems. An obvious and desirable further architectural step would be the integration of the two mechanisms to provide a single client interface to all types of transfer. A particularly intriguing possibility is that the 'by reference' mode of operation might be used to set up a real-time dialogue.

7.5 Authentication

In many applications of both computing and communication, questions of security are very important. In the context of communication, it is usually a requirement that information of a private or valuable nature is conveyed only to the intended recipient (or recipients). Additionally, the recipient generally requires assurance that the information originates from the advertised source. Furthermore, it is necessary to ensure that the information has not been tampered with during transmission. A number of these requirements, for example, apply to practical versions of the systems for monitoring and file distribution described here; another example arises in the context of the stored messaging system described in Chapter 8. There are many ways, both within the communicating end-systems and along the transmission path, in which network security may be compromised. These range from physical tapping of the transmission media, through logical diversion of information streams, and masquerading by unauthorized entities as authorized; in each case, the purposes can include unauthorized gathering of information, malicious alteration of information, and production of misinformation. A detailed review of ways in which network security may be compromised may be found in Voydock and Kent (1983). The problem of providing defences against such intrusions may be tackled at a number of levels, which range from the physical environment and hardware to the end-to-end system architecture. Substantial contributions to meeting the overall requirements can be made if ways can be found to render information unintelligible to an outsider (this may also assist in rendering attempts to introduce misinformation abortive) and to prevent masquerading.

A number of the same issues arise in the design of a computer operating system. However, since such a system resides within a single computer, solutions to security issues frequently employ mechanisms peculiar to this context and have no obvious generalization to computer networking. A notable exception to this is exemplified by the study and design of capability-based operating systems (Wilkes and Needham, 1979; Needham and Walker, 1977; Fabry, 1974; Dennis and Van Horn, 1966). Under the topic of distributed computing systems design, the security aspects of both operating systems and computer networking merge (Nessett, 1983). Since the UNIVERSE network laid claims to being suitable both for communication among autonomous computer systems and as the basis for the construction of distributed systems whose components could be widely separated, no set of experiments relating to the network could be considered complete without attention being paid to the subject of security.

Within the project, two topics which contribute to satisfying the requirements of security mentioned above were selected for study: these were authentication and encryption. The former is the subject of this section; the latter is dealt with in Chapter 9. Two of the participants in the project had a particular interest in authentication: Cambridge and GEC. It is upon the authentication experiment conducted by these participants in collaboration that this section reports.

7.5.1 *Authentication in the Cambridge distributed system*

At the time of the project, there was already in existence at Cambridge an object representation and authentication subsystem (Girling, 1982 and 1983; Needham and Herbert, 1982) which had been integrated into various aspects of the model distributed computing system (Section 6.3) in operation there. This subsystem was itself distributed and so, in addition to its primary role of providing the vehicle for studying one model of the way in which authentication might be provided in a network such as UNIVERSE, it also realized a further example of a widely distributed system operating across the network.

At the basis of the authentication mechanism is a system of object representation. More specifically, the representation of an object is defined to consist of a name, together with the identity of an issuing authority, and a means of verifying the assertion attributed to the identified authority that the name does indeed identify the object in question. Thus it is intrinsic to the representation mechanism that its entries for individual objects are susceptible of verification and it is this fundamental authentication property which enables higher levels of authentication to be built up for specific areas of application.

A self-evidently necessary feature of any satisfactory scheme for authenticatable representation is that those not authorized to hold a representation for a given object cannot steal or guess it. Prevention of unauthorized copying is not addressed here, though encryption may be used to reduce this form of system abuse. To prevent unauthorized manufacture of valid representations, an unforgeable token, called a ***temporary unique identifier*** or TUID, is incorporated. Thus the representation of an object named N, under a naming authority whose identity is A, is composed fundamentally of the tuple

$$\langle t, A, N \rangle ,$$

where token t is the TUID assigned by the representation generating service. A, the naming authority identity, is also referred to as the ***authentity***. Both A and N are forms of identifier called ***permanent unique identifiers*** or PUIDs. In fact a representation is considered essentially ephemeral, in that it should last only as long as the object to which it refers has relevance to the system. So, each such entry also has an associated time-out, δ, at the expiry of which the representation is deleted. This serves two purposes. Firstly, it is necessary to be able to ensure that representations for objects which no longer exist can be deleted and, moreover, it is important that this happen automatically if the representation falls into disuse, in order to prevent both accidental and malicious abuse of an invalid representation. Secondly, as a practical matter for the service providing the implementation of the representations, it is necessary that it be protected from accumulating useless entries to which all reference has been lost. Provision is made for maintaining the time-out under suitable authority, either to reset the time-out to a new, positive value, thus maintaining the lifetime of the validity of the representation; or to set the time-out to zero, so deleting the representation.

Reference has just been made to 'suitable authority' for maintaining a representation. At the time of requesting generation of a new representation, evidence of authority to do this must be presented. While such authority is deemed sufficient to maintain or delete a representation, it is not considered necessary (indeed, it would be generally undesirable to have to pass on such evidence to all those merely having need to maintain representations) and so a further unforgeable token, d, is added to complete a representation. This token can be regarded as a representation identifier; it is called a ***temporary permanent unique identifier*** or TPUID and has the same lifetime as the representation with which it is associated. In terms of implementation, a TPUID is the same as a TUID; however, some care is taken to ensure that, for a given representation, one is not deducible from the other and that the two together do not lend assistance to attempts at unauthorized representation manufacture.

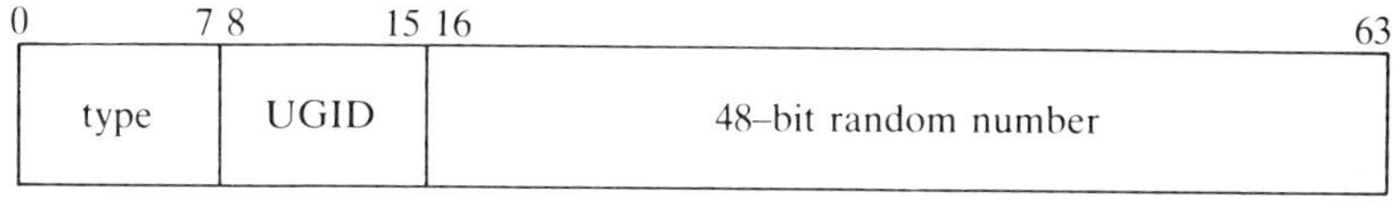

FIG. 7.7 UID format.

Thus, a complete representation consists of the tuple

$$\langle d, t, A, N, \delta \rangle .$$

Currently valid representations are stored in a table known as the *active object table* or AOT. This table is in principle global to the whole network. Its implementation may be either central or distributed. Access to this table is provided by one or more AOT manager services. Of the various services provided by the AOT managers, the following three are fundamental:

(a) Verify the representation, given t, A, and N.

(b) Create a representation, given A, N, and δ; return d and t.

(c) Change the time-out associated with a representation to δ, given d, t, A, and N.

In the Cambridge implementation, PUIDs and TUIDs (collectively called *unique identifiers* or UIDs) have similar representations (Fig. 7.7). Each is a 64-bit field, subdivided into an 8-bit type integer, an 8-bit field called an UGID which uniquely identifies the service which generated the UID, and a 48-bit random number. Type 255 is a PUID. Types 1 to 254 are TUIDs, the particular value indicating in which AOT manager service the TUID resides. Type 0 represents the null or 'impossible' token, found convenient for implementation purposes. The 48-bit random number has different significance for PUIDs and TUIDs. For PUIDs, it represents a particular solution to the problem of name allocation. TUIDs are supposed to be unforgeable and use of a sparsely allocated set of random numbers provides this to some extent. In either case, the presence of the UGID field prevents naming or token collisions when more than one service is issuing PUIDs or TUIDs.

7.5.2 *Authentication in UNIVERSE*

In the context of the original authentication scheme, there was a single site, a single AOT manager, and, conceptually, a single source of authority for naming. Moreover, the possibility of the management of the authentication system extending beyind a single site did not arise.

In expanding the authentication implementation to cover two sites, all of these premises were altered. Both sites required the ability to retain complete autonomy in respect of its authority to name its own entities. Moreover, neither site would find it tolerable to be dependent upon the other for routine access to its own resources. Thus, at least two AOT managers were needed, one to reside at each site. Furthermore, while either site would want to be able to grant access to some of its resources to the other site, neither would wish to cede control and management of its 'domain of authentication' to the other site. Generally, these requirements were considered quite realistic for two autonomously managed sites needing to collaborate on some endeavour.

To demonstrate the feasibility of extending the scheme to more than one site, two applications of the authentication mechanism were chosen: privilege management and user authentication. As an abstraction, the essential quality of a privilege is that it is conferred upon a restricted set of recipients who thereby gain permission to perform some action. The representation system described here can be used to represent various privileges. By constructing a table in which the names (authentity and PUID) of recipients are associated with the privileges they are allowed, and embedding this table in a server process which is imbued with the authority to create representations for privileges, a privilege management service may be created. Such a privilege manager can offer two services: simple checking that a given object has a particular privilege or granting (a representation of) a privilege, upon presentation of a valid representation of an object, including proof of ownership (TPUID as well as TUID), together with the PUID of the relevant privilege. An example of particular relevance is the privilege required to manage the registration of users and their passwords.

A user authenticator can be constructed in an analagous way to the privilege manager. In this case, the process has a table of user names (PUIDs) and passwords, together with the authority to create representations of users. As in the case of the privilege manager, the user authenticator offers two services. On presentation of a user's identity and password, the authenticator either confirms or denies the existence of a matching entry in its table. This provides a simple password checking service. The full user authentication service includes this function but, in the case that the check is successful, it also returns a representation of the user. Typical applications of the user authentication service include the use of terminal access for login to a remote system ('remote login') and file transfer between systems.

7.5.3 *Implementation and assessment*

The original AOT manager was implemented at Cambridge in one of the Z80 small server systems (Chapter 6). Several related services are implemented in the same server, including the privilege manager and user authenticator mentioned above. The implementations, however, make no assumptions about where each service is located and access is made entirely through named SSP services.

As a first step to providing such services at another site, a second AOT manager, user authenticator, and privilege manager were made available at GEC in another Z80 small server. Naturally, facilities for using various aspects of the authentication system had been developed at Cambridge and a number of these were available in the TRIPOS system used for the processor bank there. These facilities were made available for experimental purposes at GEC on the TRIPOS-based, 68000 large server system.

It has been noted in Section 7.5.1 that within every UID, whether permanent or temporary, there is an identification of the issuing system. This in effect was used to implement the concept of 'domain of authentication', whereby the various services associated with a given AOT only recognized UIDs issued by the same AOT manager. More general schemes, in which, for example, AOT services regard a subset of AOT managers as trustworthy, are doubtless possible but they were not investigated in the project. A particular consequence of the domain of authentication concept was that access to privileges was protected, so that neither site, for example, could interfere with the other's passwords, nor indeed perform any other privileged operation without being specifically authorized by the site concerned.

A somewhat different situation arose in the case of access to another site's TRIPOS system. During the login procedure, TRIPOS would by default refer to its local AOT services to authenticate and obtain a representation for the user. An option was added to the TRIPOS login so that it could use the user authenticator at the user's own site. Thus a user could login to any system at the user's home site or on a system at a remote site provided that the remote system considered the user's site trustworthy and the authentication succeeded.

Authenticated file transfer made use of an authenticated version of the byte stream protocol (BSP) mentioned in Chapter 6. The philosophy underlying this particular form of authenticated communication is that the initiator of the communication is the party responsible for all consequences and must therefore authenticate itself to the responding service. This it does by passing a representation of itself (or its user) in the OPEN block. Authentication of the responding service by address is presumed to be sufficiently good for the purposes of the initiator. Upon receipt of such an OPEN, the service should check that it trusts both the issuing AOT manager and the authentity of the representation and then it should verify the representation with an AOT service familiar with the issuing AOT manager. If all is well, an OPENACK indicating acceptance of the connection is returned and the file transfer proceeds. If some or all of the checks fail, the connection may be refused altogether or access to a public part of the filestore may be allowed, depending on the severity of the authentication failure and the nature of the service.

Since the authentication process is confined to the connection establishment phase, it is in principle applicable to any communication based on use of an LVC. Indeed, it is also applicable to SSP and datagram communication, though in these cases, depending upon the nature of the application, there might be questions about both performance and security.

In practice, only one other service using this form of authentication was used in the project. This was an alternative form of terminal access, where a user was already logged into one system, which consequently already possessed a representation of the user. In this case the user might use authenticated BSP for the terminal connection and the responding system performed the same checks as before. The end result was that the user was authentically logged into more than one system but had only been obliged to enter name and password once.

In assessing the overall success of such an authentication scheme, it has to be noted that various entities in the system need to be trusted. The security of the underlying communication system can be improved by using encryption; the considerable reliance placed upon the integrity of the nameserver remains. Passing a representation to another system implies trust in that system not abusing the representation and is a fundamental aspect of the system described here. A problem associated with the authenticated BSP used in the project is that the initiating system passes a representation to a system that is not authenticated. A scheme which generalizes the one-way authentication to two-way authentication is described in Girling (1983). Another shortcoming of the system is the lack of any way to check the identity of the remote authentication services; again, Girling (1983) describes an extension to overcome this.

Two practical issues need to be addressed in any implementation of the system. The first and more fundamental is that the table of current representations needs to reside in stable storage, which was provided in the project by using local fileservers. Since the AOT manager needs regular access to this storage to record changes to the AOT (including time-outs), the system is vulnerable to failure of this component. This could have been improved by use of dedicated storage, perhaps battery-backed RAM. The second feature is that PUIDs are not a user-friendly form of name and this necessitates a variety of mappings to support the interfaces to the various users of the services.

Although the process of authentication was slowed down considerably by the insertion of the satellite link, the performance of the system was considered satisfactory owing to two factors: the operations to which the authentication was applied were either of the sort which were performed rather infrequently, such as user registration or privilege management, or once at the beginning of a longer session or operation, such as a terminal session or file transfer.

Notwithstanding the various detailed criticisms of the scheme as implemented in the project, the system demonstrated the ability to provide authenticated communication within and between sites, each of which retained control of its own authentication domain. Extension of the original, single-site implementation to two sites was found not to be difficult, owing partly to the segmented design of the original system and to its use of idempotent SSP, a protocol which was directly supported by the network.

7.6 UNIVERSE Support Environment

A subject of computing technology which has received a great deal of attention is software engineering, the process of software design, development, and maintenance. In its most primitive form, omitting the design aspect, the program development cycle involves using an editor on program source files to produce a new version of source. A compiler is then used to produce modified versions of object files; listings may be produced at various stages of the process. Depending on the nature and complexity of the software, the object code will be executed in some environment varying from standard execution, through an interactive debugger, to specialized environments for real-time or other special purpose testing. To back up such a development cycle there needs to be at least a methodology, and preferably some automated assistance, for keeping track of the numerous versions and modifications of the software. This becomes indispensable when the software is being developed by a team rather than one person. More recent efforts have included work aimed at replacing the 'editor' with a more sophisticated interactive system which presents a structured view of the code being generated. It operates in conjunction with a range of tools which not only prevent invalid use of the programming language but also help to link the organization of the code and data structures with an abstract representation of the system design. Any such system would be expected to operate with the capability to access a range of consistent versions of the design and corresponding software.

In the context of a consultancy developing software for a variety of clients, owing to the diversity of target systems, a wide range of software tools is required. Even within a single project, it is quite likely that not all of the data, resources, and tools will reside on a single system; also, the target system environment may well be quite different from that of the development system. Stimulated by these considerations, Logica proposed, and undertook with GEC and Cambridge, an investigation of a distributed architecture for software development. This architecture and its realization as part of the project was called the *UNIVERSE Support Environment* or USE experiment. The objectives were to observe the effects of the network delays on various aspects of the development cycle and to investigate the network facilities needed to implement such a distributed facility. For this purpose, the model taken of the software development process was less sophisticated than the advanced one outlined above; it was, effectively, the fundamental model, enhanced by the addition of elements for project definition and control.

7.6.1 *The framework*

At the heart of the architecture for the distributed environment for software development is the notion of a relatively static framework, to which users and resources may be attached dynamically. Within the framework are two managers: a resource manager and a database manager. The latter maintains information on a range of objects at several levels. Generally, the information is arranged according to a set of schema appropriate to the concept of a 'project'. An example is the team structure of a project, specified in terms of the names of people and their roles within the project, such as project manager, programmer, librarian, etc. Documentation is classified under such headings as operational requirement, functional specification, systems design, module design, user reference, and so on. Members of the team are recorded, together with their roles and relationships to documentation, code, design, etc.

The database also maintains a record of the structure of the software, including relationships and dependencies among modules and the state of various versions of program source, object, and listings for individual modules. Responsibilities of members of the project team for the various modules is recorded, together with the corresponding access rights. One service offered by the system enabled users to query the database for these details about the project. More central to the USE system operation was the support provided by the database manager for organized storage of project software. The database manager contains references to these data files, upon which the various tools will be called upon to operate. The files themselves may reside in any convenient filestore, under the access control of the database manager. Requests to access segments of code, whether source, listing, or object, resolve to references to a file or concatenated set of files from external file storage. Using a private naming convention, the database manager accesses these, retrieves and concatenates the files, and delivers the resulting unit to the requesting client component.

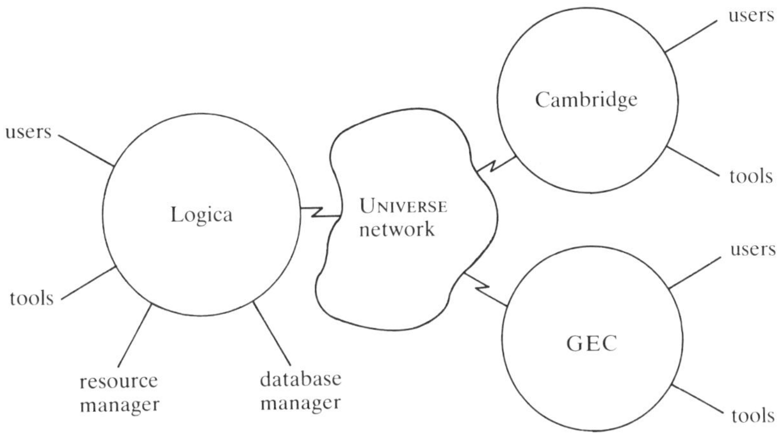

FIG. 7.8 Disposition of USE elements.

The dynamic attachment of software tools (resources) and users is organized through the resource manager. Resource services, such as compilers, debuggers, and printers notify the resource manager when they are available for work. Users also make their requirements known to the resource manager through a system stub known as a *user interface machine* or UIM. The resource manager then acts as a broker, matching requests to available resources. The 'introduction' having been effected, the resource manager plays no further role in the performance of whatever function was requested. It simply notes that the resource is busy and reverts to waiting for further notifications of available resources and further requests from users.

7.6.2 *Implementation*

The elements of this model software engineering development environment consist of the framework, resources or tools, users of tools, and data; the latter consist of programs, both source and binary, and information on project definition and control. The framework contains the internal organizational elements of the system, namely the resource and database managers. The disposition of these various elements among the sites collaborating in the experiment is shown in Fig. 7.8. It will be seen from this that the whole framework was situated at Logica. There were also users and tools at Logica, so that the site could function independently as a self-contained realization of the USE concept. At GEC and Cambridge, tools were made

TABLE 7.2 USE system components

System	*Site*	*Component*
Prime 550, PRIMOS	Logica	RAPPORT nucleus and database
PDP 11/44, XENIX	Logica	Filestore Database manager Resource manager C compile server Print server XENIX execution server UIM
GEC 4065, OS4000	GEC	AIDA debug server OS4000 execution server PAD (access to remote UIM)
Large server, TRIPOS	Logica	BCPL compile server TRIPOS execution server UIM
Large server, TRIPOS	Cambridge	BCPL compile server TRIPOS execution server UIM

available for remote connection to the framework and there were also users of the system; these sites could not function as part of USE without Logica.

A number of different computer systems was available at the sites participating in USE. From the various software development tools indigenous to these systems, a representative set was chosen for incorporation into USE. Table 7.2 lists the various components of the USE implementation according to site and computer system.

The USE database manager was distributed over two systems at Logica. The Logica general-purpose database system (RAPPORT), on the Prime 550, provided basic database support and the special-purpose USE database manager was implemented under XENIX on the PDP 11/44. The XENIX filestore was also used to hold all software files of program source, object code, and listings.

The resource manager was also implemented under XENIX on the PDP 11/44 and included the following functions:

- maintaining lists of available resources;
- managing queues of users waiting for busy resources;

- allocating resources to satisfy requests from other components of the system; and
- accumulating usage statistics.

When a component requesting service of a resource is matched with an instance of the relevant resource, details of the requested service, including its origin, are passed to the resource; the address of the allocated resource is also passed back to the client component; thereafter, the client and server components communicate directly. The resource manager records the lists of resources and queues of waiting users in a database managed by the RAPPORT system. This database is logically quite distinct from that containing the project definition and code files.

The UIM is implemented at the *command line interpreter* (CLI) level, an approach which offers a number of advantages in an experimental situation. The incorporation of new facilities amounts to the introduction of a new command, or possibly a set of commands. Implementation by means of a set of command files not only enables the execution of intermediate programs in a fashion invisible to the user but also makes the UIM definition quite easy to transport. As can be seen from Table 7.3, UIMs were implemented in both TRIPOS and XENIX.

Special programs or 'stubs' were created to provide commands for incorporation into the UIM to interface to the distributed parts of USE, such as the resource and database managers and, indirectly, to those tools accessible to the user, typically compiler, printer, execution, and debug services. These program stubs accessing the remote servers were termed 'ghost' servers; the general mode of operation of such 'ghosts' was to contact the remote service, provide it with the data to operate on, receive any resulting output, and report the results back to the user. In addition to the foregoing, the CLI approach to UIM implementation made it quite simple to incorporate access to a local resource, such as a compiler. In this case, the database and project control aspects of the USE framework operated in the normal way but the resource was regarded as a private one and was not registered with the USE resource manager. Table 7.3 shows the set of facilities available via the UIMs on the two principal operating systems on which UIMs were implemented.

On the GEC 4065 system at GEC, PAD access to remote UIMs was provided and local AIDA debug and BABBAGE compile services were also available; a full UIM, however, was not implemented for OS4000.

No firm rules were laid down about how the network services were to be used in providing communication among the constituent components of USE, it being regarded as more constructive to gain varied experience from which to draw conclusions about how to implement such a system.

TABLE 7.3 UIM facilities

System	*Facility*
TRIPOS	BCPL compiler
	print ghost (server on PDP 11/44)
	execution service
	database interrogation
	edit service
XENIX	execution service
	C compiler
	BCPL ghost (server on large server)
	AIDA ghost (server on GEC 4065)
	database interrogation
	edit service
	print service

Generally, however, communications divided into two broad categories: short transactions, for control and interrogation purposes, and more extended sessions, generally associated with the transfer of bulk data but also used in providing interactive access to remote debug services. Contrasting procedures are used by the BCPL and C compiler servers in respect of whether the UIM 'ghost' server or the remote server accesses the framework database manager to obtain the data for the compiler to process. The BCPL server is passed a suitable database reference by the UIM and makes its own access to the database; whereas, for the C server, the C 'ghost' constructs a file containing both the request and the data which it obtains from the database manager.

All short transactions for control were made using SSP as a primitive form of RPC. Most forms of more extended communication used BSP or its standardized form, CR82 (Section 6.5) as their basis, so as to provide a reliable, if slow, form of transport. For terminal communication, a primitive form of TS29 (PSS User Forum, 1981) was used. For file transfer, several protocols were used, by mutual agreement among the implementors concerned. These included the remote file access protocol defined in one of the other experiments (Chapter 9), TRIPOS transfer utilities, and an *ad hoc* method based on SSP.

7.6.3 *Conclusions*

This experiment was a considerable undertaking. Conceptually, it was broader than any of the other systems described in this chapter, as could be expected of a project having an end-user system as goal. Furthermore, the implementation was required to cope with a diversity of systems. Though there had been earlier consideration of a less ambitious version, the eventual form of the experiment reported here began in mid-1983; consequently, the system reached the stage of being ready in the latter part of 1983, in the last months in which the OTS was available to the project. Most unfortunately, this coincided with a time when the earth station at UCL became unreliable, thus almost severing Logica from its collaborators in USE.

A consequence of the network unreliability was that the robustness of the underlying USE system and the procedures for communicating among the components were tested rather thoroughly. The BSP-based protocols could survive quite a long hiatus in communication; through the use of idempotency and time-outs, the SSP-based interactions also achieved considerable resilience. Throughout the experiment, neither of the constituent framework managers became corrupt through interrupted or prematurely terminated transactions. In achieving this, the design decision of having the resource manager behave in an essentially passive fashion, mediating between active users requesting resources or tools which have actively made themselves available for work, is considered to have been an important contributing factor.

A further consequence of the unreliable state of the network during the test phase of USE was that since nearly every message passed over the wide area segment had to be repeated, possibly triggered by time-out, response was very poor. This was further exacerbated by USE having to share both the PDP 11/44 and the Prime 550 multiaccess systems with other, unrelated activities which imposed significant loads on both machines. It says much for the system that it reached the stage of being used to maintain its own project and code. However, it was clear that performance would require substantial attention. While some aspects of bad performance were clearly the result of network problems, overloading of particular resources, and use of relatively slow mechanisms for bulk data transfer, it was not possible to decide if there were fundamental aspects of the design which might also contribute to the poor performance.

This experiment can be regarded as having demonstrated the feasibility of the concept of a particular open architecture for a programming support environment catering for the dynamic attachment of an essentially unlimited range of specialized software engineering resources. Conceptually, the framework may be regarded as the vehicle for providing explicit interfaces for the attachment of both resources and users. Several further developments of the basic architecture immediately suggest themselves and could be relevant for applications in a variety of circumstances. Examples are the inclusion of authentication and security and incorporation into the system of so-called 'fourth-generation language' facilities for a variety of applications. For the usual reasons of increased capacity, resilience, and flexibility, it might be desirable to develop distributed versions of the framework components. Some of these developments, such as distributed database systems, are of course substantial areas of active research in their own right. Another aspect implicit in the whole approach is that if the concept of the framework reached the stage of maturity where an operational system could be based upon it, then it would also be a strong candidate for standardization in order to allow a wide range of resources and end-user workstation systems to attach to it.

Generally, the project demonstrated the feasibility of both concept and design approach. A prerequisite for the further development of any system such as this to operational or product state is the ubiquitous availability of suitable wide area, high-speed communications infrastructure. To have reached this stage, in the face of considerable technical obstacles, constituted an achievement upon which the participants were to be congratulated.

7.7 Summary

Distributed systems design techniques applied to the construction of systems confined to regions of a size typically spanned by a single LAN may be regarded simply as offering an alternative design methodology to that exemplified by the use of a single, centralized CPU. At least in principle, the choice between the two approaches may be made by assessing the advantages and disadvantages of each in the context of the requirements of a particular system or application.

The investigations into distributed systems described in this chapter have a different emphasis. The geographical dispersal of the component parts compels a distributed approach. The question is whether techniques, such as SSP or RPC, developed for use in a confined area can be extended for use in a wider area

The way in which functions were distributed in the systems described here varied in nature and degree. In the remote fileserver and USE experiments, different functions, each essential to the overall system operation, were provided at different sites; these systems could not function in the absence of full network connectivity. In the other examples (monitoring, file distribution, and authentication), the emphasis was on replicating elements of systems, so as to provide wider and more flexible coverage for a service than could sensibly be provided by a single server and, in the case of monitoring and file distribution, so as to mitigate the effects of intermittent network connectivity.

The use of SSP was important in all of the distributed systems experiments, for control interactions, the transfer of short bursts of data, and establishing synchronization. The incorporation of SSP into the set of network access protocols enabled applications which had formerly worked on a single LAN to operate with little or no change over the UNIVERSE network, the adjustment most frequently required being the length of time-outs. In terms of the success or otherwise of distributed systems 'transplanted' from a single LAN to the UNIVERSE network, applications could be roughly divided into three categories: those for which time (absence of delay) was of the essence, those in which low delay was important but the application was amenable of some degree of compensatory re-engineering, and those which were largely insensitive to delay. An example of the first is screen editing from a remote terminal: delay of the order of that imposed by a satellite link renders such activity essentially unworkable for serious use. At the other extreme, an activity such as compiling source from a remote filestore and storing the result as a file on the same (or any other remote) store is an activity relatively insensitive to delay: it is much more important that the network throughput match the rate of compilation. The remote fileserver experiment was an example of a system where low delay was important but where re-engineering, in the form of increased use of cacheing, could be used to hide many of the effects of the long network delay. In the USE system, both high throughput and low delay would be necessary.

The purpose of the distributed systems experiments was to focus on the use of rapid, transaction-style interactions over a wide area network, rather than throughput issues. Clearly, in absolute terms, the long delay introduced by the satellite segment was a hindrance. However, the simplicity of the SSP mechanism, coupled with the use of cacheing in the nameserver to avoid repeated long communication-setup interactions across the wide area network, enabled relatively closely coupled distributed systems to be constructed and demonstrated. With the use of forward error correction in the satellite segment, the reliability of the SSP mechanism was sufficient for error detection by time-out and recovery by idempotent repetition at the application level to be satisfactory because they were seldom invoked under normal conditions. However, experience with a long-delay network demonstrated in an exaggerated way several problems intrinsic in the use of time-outs as a method of detecting errors and triggering recovery (see also Section 3.7.2). The issue is one which affects not only transactions but also bulk transport: the effect on the latter, and a suggested remedy, are described in Section 9.7. The problems for transactions arise from the use of a single end-to-end time-out to trigger further action in case of no reply from a remote service. Among other factors, this time-out needs to take into account the 'distance' (in terms of the number of network segments and switching elements and their respective delay properties) separating client and server. Though a palliative was applied, in the form of a flag set by the nameserver indicating that the service was 'remote' (not on the same LAN as the client), the problem was not solved within the project. The simple, synchronous (blocking) approach to the use of SSP generally taken in a single LAN environment served perfectly adequately in a multiple LAN network with long delay. It served to emphasize the advantages of the 'thread' concept, whereby complex operations are broken down into a set of parallel courses of sequential action and these threads are then scheduled in conventional fashion. This methodology is of course intrinsic to RPC mechanisms.

In summary, the incorporation of SSP as a network protocol enabled the successful construction and demonstration of a representative sample of wide area distributed systems. This achievement provides substantive support for the idea that it is possible to build wide area networks which have many of the properties of LANs and which support similar styles of access and use.

8. Voice and images

8.1 Multi-media services

The traditional communications services of telephony, telex, and radio and television broadcasting have been and are currently carried by separate transmission networks. These networks convey signals in analogue form and in the main use dedicated channels or circuits for single connections. The ability to digitize these signals, particularly those of voice and video, in an economic fashion has opened up the possibility of building networks that carry multiple services, that is, combinations of the traditional communication services and computer data. Such combinations, including those associated with computing, are now often referred to as 'multi-media' services. Since computer data is normally carried in packet form, there are three ways in which a multi-media services network can be constructed. A circuit-switched technology may be used, in which case methods of networking computers must be adapted; packet-switched technology may serve, in which case adaptation of the traditional communications services, such as voice, video, and facsimile is required; or some form of hybrid network carrier must be constructed. This chapter is concerned with the relatively little-investigated part of the provision of multi-media services in a packet regime, namely, the transmission of digital voice and video signals. Specifically, it describes the work undertaken to provide voice and image transmission services over the UNIVERSE network.

Voice and video signals are converted to digital form by sampling the signal at regular intervals and digitizing the signal amplitude at these sampling times. For voice the standard method is to sample 8000 times per second and to digitize the sound amplitude with eight bits giving a constant 64 kbit/s digital stream. For video the sampling rate is much slower at 25 times per second, but the information sampled is much larger, being the light amplitudes for red, green, and blue at each of a two-dimensional grid of pixels. Without some form of compression, good quality colour video gives very high digital signal rates of the order of 100 Mbit/s. Clearly digital voice and video can be carried on dedicated circuits provided that the bit rates allowed are high enough. No timing information is carried with the sampled information, constant frequency and delay being relied upon for accurate reproduction of the signal at the receiving end. This type of information, in packetized form, may also be carried across packet networks.

However, in this case transmission delay across the network will not be constant, since, although packets are generated at a constant rate, delivery is subject to statistical fluctuations caused by the shared use of channels and the use of store-and-forward elements, such as bridges, in this type of network. Although these fluctuations are not normally significant for the transmission of computer data, they must be strictly controlled if voice and video are to be carried satisfactorily. This is particularly the case for voice where the sampling rate is very high.

These considerations may be summarized in the two following, somewhat contradictory requirements for an integrated, packet-based network carrying voice and video together with computer data.

- Computer data require some means for ensuring error-free communication across the network, but do not generally need tight constraints on packet transit times.

- Voice and video data require tight constraints on transit time 'jitter', but can accept some loss or corruption of the information during transmission.

An important goal of the project was to assess the potential of the network architecture, particularly the LVC, as the basis of a packetized integrated network. The performance of the architecture in relation to computer data is the subject of Chapters 7 and 9: here we are concerned with real-time traffic. The basic transmission speeds of the UNIVERSE network were quite adequate for carrying digital voice traffic, and protocols for voice communication on the network were designed and implemented. Video could not be carried on the network without very heavy compression and equipment for this purpose was not available to the project. However, an implementation was made of services for transporting video 'stills' and slow-scan video, that is video with a frame sampling rate of one frame every few seconds. Some theoretical work on compression, for both still and moving video, was also undertaken.

The packet voice system was built around an adaptive protocol, which adjusted to the prevailing network conditions of delay and jitter, perceived in the course of each individual call. The protocol was used for both 'real-time' telephone connections between sites and for a store-and-forward voice messaging system. The use of packets makes it easy to suppress the transmission of voice data during periods of silence. Hardware for silence detection was implemented and advantage of this was taken in the protocol design. Details of the protocol and its implementation, together with a number of performance measurements are described in the next two sections.

Work on the transmission of video centred on the use of microprocessor-controlled framestores which were interfaced to the network. Framestores are able to capture single images in real time from a TV camera and to refresh TV monitors at their normal frame rate from a memory. Thus they allow video transmission activities to be isolated from the video sampling rate of 25 frames per second. At the extreme they can be used to capture a single TV frame and to transmit this image to a corresponding framestore which will keep the 'still' image displayed on a TV screen. A number of framestores were built for experimental use in the project. These were used for the transmission of slow-scan video and video stills both in a store-and-forward and a real-time mode. This work is described in Section 8.3.

8.2 Packet voice

8.2.1 *General considerations*

In circuit-switched networks, the bandwidth and delay associated with a call are both constant for the duration of the call. Thus circuits provide a very suitable medium for voice conversations at the cost of some inefficiency in the use of bandwidth. Bandwidth on a circuit is always available when it is needed, but in general, for a conversation, use of this bandwidth will be required for less than 50 per cent of the time; during the remaining time 'silence' will be being transmitted. In situations where circuit costs are high, such as in transcontinental links, the time-assigned speech interpolation (TASI) scheme (Bullington and Fraser, 1959) has been devised to make more efficient use of circuit bandwidth. In this scheme a number of voice channels share a smaller number of circuits through voice-activated switching. Each voice channel alternates at random intervals between 'talkspurts' (i.e. periods of continuous speech) and silence, and is only assigned a circuit during talkspurts. A channel is given a circuit upon initiation of a talkspurt and releases the circuit when the talkspurt is completed. If all circuits are busy when a talkspurt begins, speech is lost until a circuit becomes free.

In packet-switched networks, bandwidth is only claimed when it is actually required and thus, in principle, more efficient use can be made of the available transmission capacity. This advantage is somewhat offset by the need to carry addressing information with each packet, but more importantly there is the penalty that the transmission delay through the network is not constant. In packet-switched networks, calls share the use of communications channels, this sharing being effected by store-and-forward components in the network which, if necessary queue packets for transmission on the shared channels. This queueing produces a variable delay for the transport of the individual packets of a call across the network. This variability of delay has two components, a continual statistical jitter and longer-term drift in the average delay caused by changes in the level of traffic load on the network. The amplitude of the jitter generally increases as the average delay increases. The principal function of packet voice protocols is to deal with this variable transmission delay.

A telephone conversation is clearly a duplex activity, requiring separate channels in each direction between speakers. Normally, computer networking protocols are also duplex, not only in that data may flow in both directions but also in the sense that data flowing in one direction is controlled by protocol messages flowing in the other direction. In contrast, packet voice protocols are simplex as no control information flows back from the receiver to the transmitter. Since voice data is generated at a fixed rate, flow control mechanisms are not appropriate, and any attempt to correct transmission errors by requesting retransmission introduces the possibility of massive extra variability in the delay. Thus the two halves of a telephone conversation can be treated for packet protocol purposes as different calls. This has the consequent benefit that the same protocol can be used in voice messaging systems.

In circuit-based voice transmission, the receiver's timing is normally data driven, that is the receiver plays back the data as soon as it is received. Since the delay caused by a circuit is constant, this means that the receiver's timing is controlled by the transmitter. In the packet case, the receiver obviously cannot derive its timing in this way and must control its own play-back timing. When the first packet on a call reaches the receiver, it waits for a short time interval and then begins its play-back. Time in the receiver is measured from the moment when this play-back starts; the time interval before play-back serves to allow for some degree of time jitter in the network delay. The receiver plays back subsequent packets at the appropriate timing required to achieve the agreed coding rate. Normally, packets will arrive before they are required for play-back, but on occasions a packet may be subject to a large delay and arrive too late to be played back. In the simplest protocol, a receiver would stay rigidly with its elapsed timing count and discard any samples that arrived too late for play-back. In the more complex protocol developed in the project, continued late arrival of blocks is taken as an indication that the network load has increased, and that the timing of the play-back should be 'retarded' to avoid fragmentation of the sound. This extra retarding of the play-back should continue for so long as the conditions of heavy loading producing the longer average delay and jitter persist. A matching 'advance' mechanism for reducing the delay as the network load subsides must also be provided.

In a packet voice protocol it is a straightforward matter to suppress the transmission of voice samples during periods of silence. Each packet is given a sequence number and packets that are full of sound samples representing silence are not transmitted. The receiver uses the sequence numbers to detect the absence of packets and to play back the next packet received at the correct time. Packets that are lost or corrupted in transmission will, in this case, be treated automatically as silence. The protocol developed in the project accommodated this type of 'silence transmission' suppression.

8.2.2 *The voice protocol*

The connection of two phones in a 'conversational' manner will be assumed in the following description of the protocol. The protocol may also be applied to message systems for stored voice, but in this case the timing constraints are much less severe.

Transmission is quite straightforward. Once a connection has been established, the transmitter begins to assemble voice samples. The voice is sampled at regular intervals of $d\,\mu$s, where $d = 125$ for the standard sampling rate of 8000 samples per second, and assembled into fixed length packets containing s samples. When packets are complete, a sequence number is added, the sequence number being incremented every time a packet is assembled. The transmitter then checks to see if the packet consists entirely of samples representing silence. If this is not the case, the packet is transmitted, otherwise it is discarded unless it is the first packet to be generated, in which case it must be sent to allow the receiver to start its timing. Note that, since the sequence number is incremented every time a packet is generated, regardless of whether it is actually sent or not, the sequence number will be a linear function of time. The receiver will thus be able to stay synchronized, detecting missing packets by gaps in the sequence numbers. Some maximum silence period will normally be defined after which the transmitter must send a packet even if this only contains silence samples. This serves both to prevent the network timing out the call and to indicate to the receiver that there has been no loss of connection.

Reception is a more complex process. After the first packet has been received and the prescribed delay time waited, play-back is started. At any given time during play-back, the receiver will have a reserve, say r, of voice samples already received and waiting to be played. The value of r at the time a new packet arrives, called R, gives an indication of the current play-back delay in the receiver. If a packet arrives with a sequence number indicating that the b preceding packets are missing, then $b \times s$ silence samples must be added to give a reserve of $R = r + b \times s$ before the new packet is added. During a long period of silence all the reserve samples will be played back and r will count down to zero. At this point the receiver will play back silence samples and start to count r negative, that is as a deficit rather than a reserve. When the next packet arrives the reserve will be recalculated as $R = r + b \times s$, which should be a positive value for R. The new packet will then be added to R silence samples to form the current reserve.

It may happen that the addition of the product $b \times s$ to the negative reserve still leaves the reserve negative, indicating a genuine deficit in the receiver. In particular this will be the case if there are no blocks missing, and the product $b \times s$ is zero. In this circumstance the receiver has two possible courses of action, either to keep to strict timing and discard from the newly arriving samples as many samples as the value of the deficit, that is R samples, or to start immediately to play back the new samples. If the latter strategy is adapted, the receiver is effectively always increasing the play-back delay whenever a deficit incident occurs. The voice protocol developed in the project used this variable delay approach.

The strategy of playing back all samples that arrive copes automatically with the case of average delay and jitter increasing during the course of a call. In these circumstances silence periods are lengthened or extra silence periods are added. A corresponding mechanism is required to deal with the opposite circumstance of a build-up of a larger reserve than is necessary, that may come about from an improvement in loading conditions on the network. The network may also be subject to a gradual drift in mean packet delivery time which the protocol should be able to accommodate. The method used to achieve this adjustment is to inspect the values of R, the reserve when a new packet arrives, to determine the minimum value of R, say $R_{\min}$, during a given time interval. The play-back delay is then shortened by $\alpha\, d\, R_{\min}$ seconds, where α is a constant in the range $0 < \alpha < 1$. If there has been a negative R value, that is a deficit, during the interval, no shortening of the delay takes place. The timing advance necessary to reduce the delay is actually effected by shortening the next period of silence to occur. In this way the listener should not be aware that the adjustment has taken place. The procedure is repeated at regular intervals of a few seconds to provide the other half of the dynamic delay adjustment mechanism. During periods of silence there may be too few values of R during the time interval to make a significant determination of $R_{\min}$. In this case the minimum is taken over a period of two or more time intervals.

One further question that has to be dealt with is that of how the initial estimate of play-back delay is decided. This is the time interval between the first packet being received and play-back being started. Clearly the delay jitter to be expected on a call is a function of the connection path. In the UNIVERSE network architecture, knowledge of the properties of connection paths resides in the nameservers. When a name is looked up which implies the use of the voice protocol, the nameserver replies in the usual manner with an address, but also gives a characteristic of the route. The function code is used to supply a value for the initial play-back delay to be used on the call. This value is retained by the call initiator and also passed through in the OPEN block, so that both ends know the expected characteristic of the connection.

The delay value returned in the function code can additionally be made to suit the nature of the remote machine. The protocol described above was also used to implement a stored message service, in which record and play-back facilities were implemented on a minicomputer with a disk-based file store. Since this type of machine runs under an operating system and may be handling a number of different tasks at any one time, it is, in general, unable to launch packets with the regularity expected of a dedicated device. However, since there is no real-time constraint on the play-back of a stored message, a long delay between the transmission of a packet onto the network and its eventual play-back in a phone or loudspeaker is of no consequence. The flexibility of the nameserver mechanism allows a large jitter estimate to be given when the remote end of a call is a message storage system, so that 'smooth' speech quality may be obtained in spite of the irregular launching of packets onto the network.

8.2.3 *Implementation*

Standard telephone handsets were interfaced to the Cambridge ring LANs using the Z80-based small servers (see Chapter 6) in the hardware arrangement shown in Fig. 8.1. The telephone interface card, designed by GEC, contains a coding/decoding (codec) chip that handles the conversion between the analogue voice signals and 64 kbit/s PCM signals. For 64 kbit/s PCM, the voice signal is sampled 8000 times per second and the amplitude digitized using eight bits per sample. The 256 possible values of amplitude are not spread uniformly across the possible signal range but are distributed with non-uniform intervals so that most points fall in regions of amplitude that are most heavily used in speech. The particular digitization scheme used by the codec chips is the A-law quantization (Owen, 1982). The interface card is connected to the handset by a four-wire connection. Hardware on the board implements status monitoring of the telephone and silence detection. Silence detection consists of comparing the rectified signal from the handset microphone with a preset reference voltage. This arrangement cannot cope with wide variations in background sound levels, but a single level could be set for any particular room to give quite satisfactory results. Silence detection is presented to the Z80 as a status bit, allowing the Z80 to integrate silence over any required period. A timer on the processor board provides an interrupt to the Z80 every 125 μs during a call, at which times voice samples are transferred to and from the codec.

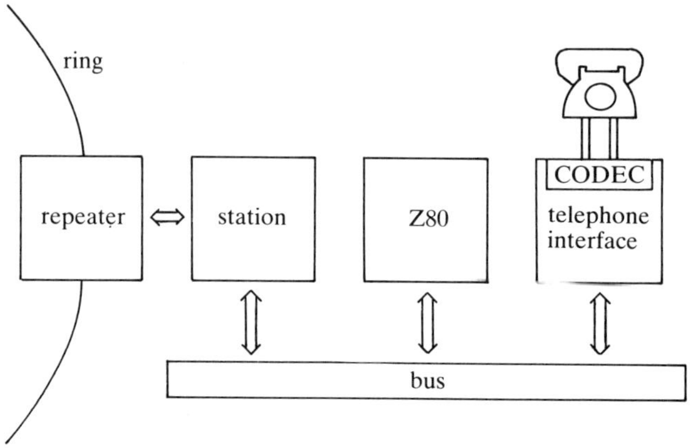

FIG. 8.1 **Telephone interface hardware arrangement.**

An important choice to be made in the implementation of the protocol is the size of packet to be used. From the point of view of the load that is imposed on the network components, large packet sizes are desirable. However, there are other considerations. Forming groups of samples into packets introduces an extra end-to-end delay of $s/8$ ms, where s is the number of samples carried in a packet. For conversations, delays above 150 ms are noticeable, but in the range 150 ms to 400 ms are considered to be acceptable; delays above 400 ms are difficult for speakers to manage (see recommendation G.114 of CCITT, 1981*c*). The satellite transmission delay makes operation in the 150 ms to 400 ms range unavoidable, but clearly the packet delay should not add significantly to this. This would indicate the use of a packet size corresponding to a delay of rather less than, say, 50 ms, that is 400 samples per packet. (In situations where telephones are interfaced to the network without echo suppression, much lower delay limits would be applicable. These limits are, however, much smaller than the satellite delay and thus not a consideration for the UNIVERSE network.) For stored voice messaging, delay is not important and large packet sizes may be used. Since the satellite bridge was severely limited in its capacity to handle large numbers of packets per second, all the experimental work on voice in the project was carried out with packet sizes of 256 samples (i.e., bytes) or more, with 1000-sample packets normally being used.

The software running in the Z80 that handles the telephone is divided into two parts, a part that deals with call setup and a part that implements the voice protocol described above which runs during the course of a call. The code is in Z80 assembly language to gain the requisite processing speed and uses a large fraction of the processing power of the Z80. It was designed and implemented by Cambridge and RAL.

TABLE 8.1 Facilities of UNIVERSE telephone service

Service type	*Facility*
0	record message for specified user
1	play back specified message
2	play back all new messages for specified user
9	set up call to specified user telephone

The handling of call setup is quite complicated, as correctly timed transitions between dialling, ringing, speaking, and various 'fault' conditions, which include such expected conditions as engaged, must be provided. Normal telephone conventions are used with standard UK tones being produced by the software. The number which a user dials to initiate a call has one of the forms

$$\langle \textit{service type} \rangle \; (\langle \textit{site} \rangle \; \langle \textit{user number} \rangle \mid \langle \textit{message number} \rangle) .$$

The *site* corresponded to the site component of a UNIVERSE network address and the *user number* denoted a particular telephone station on that site. This latter number was associated with a particular user and was usually chosen to be the same as the user's normal telephone extension number. Four types of service were available, corresponding to the values of *service type* shown in Table 8.1. After decoding the dialled number, the Z80 sets up a call in the normal manner, making use of the nameserver. A general indication of the configuration of speech equipment used for experiment and demonstration on the network is given in schematic form in Fig. 8.2. The disposition of such equipment varied during the project and included the sites at Cambridge, GEC, Loughborough, RAL, and the exhibitions.

As implied in Table 8.1 and Fig. 8.2, a voice message storage system was implemented. To record a message, the caller dialled the intended recipient's number preceded by a service type code 0 and, after hearing the acknowledging tones, spoke the message. The recipient was informed of the message by electronic mail, which included the *message number*. Two mechanisms were available for retrieval; these could be used from any telephone on the network. By means of service type code 1 and the *message number*, the user could replay any message in the store. Alternatively, the user's own number, preceded by service type code 2, could be used to provide play-back of all messages recorded for the specified recipient since she or he last accessed the store in this way. This simple service was available for experimenters and was also used in the demonstrations (Chapter 10). Within its limitations, it worked well.

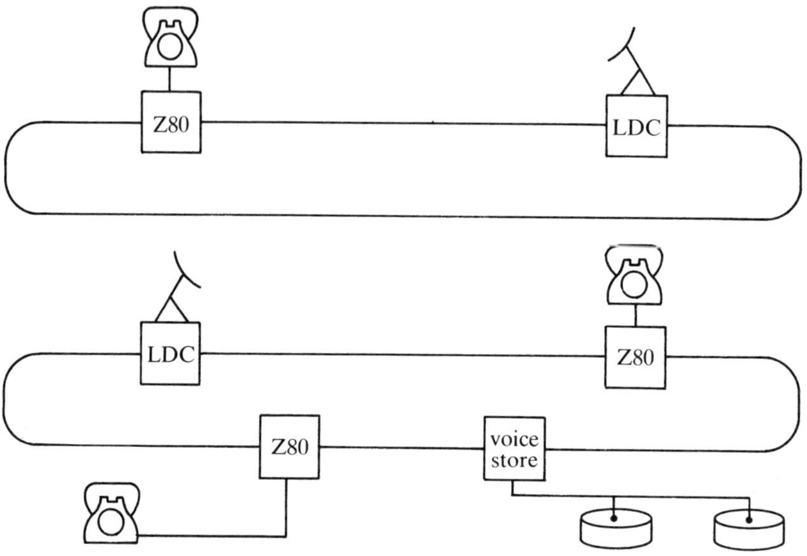

FIG. 8.2 Packet voice equipment.

In common with most telephone systems, this system exhibited the problem of identifying the user by an assumed relationship with a particular telephone number. In particular, any user could access another user's messages. In an integrated office system, a user would be expected to identify herself or himself to the system as a whole, by whatever means. Thereafter, a system of authentication, such as is described in Chapter 7, could be used to assist in ensuring, for example, that any message is only accessed by those suitably authorized.

8.2.4 *Experimental results*

The protocol was tested out on single rings, across ring–ring bridges, and across satellite connections. In the first two cases the results were very good and some typical sample monitoring data is given in Fig. 8.3 and Fig. 8.4, showing the protocol coping with particular types of network behaviour. Connections across the satellite, although giving satisfactory speech quality for most of the time, did suffer from two particular defects discussed below.

The Z80 small server was a far from optimal vehicle for monitoring the performance of the voice protocol, being near to its performance limits in simply handling the protocol. The method adopted for monitoring was to dump simple status information into an unused part of the Z80 memory as the protocol was running and subsequently to retrieve this information for analysis off-line in a larger computer system.

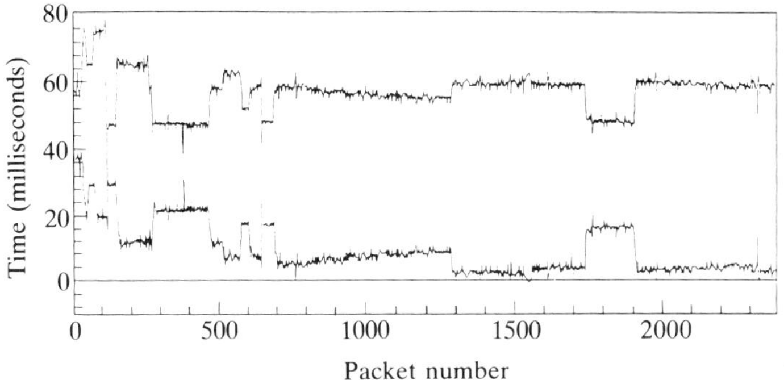

FIG. 8.3 Effect of light bridge loading. (© *1985 IEEE*)

The graphs in Fig. 8.3 and Fig. 8.4 show typical behaviour of the voice protocol for conversations between phones on two rings joined by a ring–ring bridge. The lower trace is the reserve held in the phone when each packet arrives; the upper trace shows the corresponding transit time of the packet through the network. The reserve held is an absolute quantity shown in milliseconds rather than voice samples (i.e. number of voice samples times 0.125 ms), whereas the delay through the network is a relative quantity. Since the rate of transmission of the samples is fixed and known, the relative delay can be measured by the receiving station and does not involve the measurement of the delay between the stations.

Figure 8.3 shows a typical period of running under conditions of light bridge loading. The first half of the figure shows the behaviour of the mechanism after a period of heavy load has subsided. Initially the reserve, which had been built up to cope with the load, is larger than necessary for smooth sound quality. The adjustment mechanism gradually reduces this reserve, as can be seen by comparing the reserve against some constant value of network delay, say 56 ms, as the trace progresses. In the second half of the trace, the bridge is under light rather than zero loading; periodic dips in network delay probably indicate no other use of the bridge at these instants. As can be seen from the lower trace, under these conditions the protocol does not attempt to produce a complete absence of deficits but only to keep their occurrence infrequent.

In Fig. 8.4 a sudden high load has been placed on the bridge. The level of reserve held prior to this load increase, that is the reserve at the start of the bottom trace, can be seen to be insufficient for the delays now encountered. In this case a single large deficit occurs which retards the timing sufficiently to cope with the new load. The load, which was in fact a series of file transfers, can be seen to be bursty. Because the time constant for the adjustment mechanism in the phone is quite long, it does not compensate for the reduced delivery time between bursts so

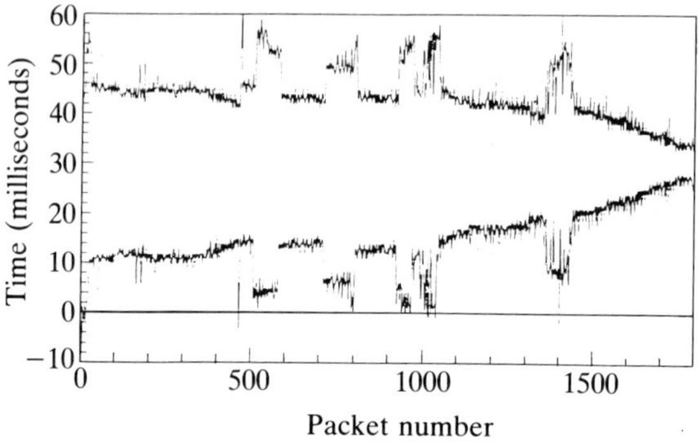

FIG. 8.4 Effect of heavy bridge loading. (© *1985 IEEE*)

rapidly as to cause another heavy deficit at the start of the next burst. At the time when this trace was produced the sound quality was reported by the listeners to be quite good. A single large deficit, particularly if it occurs within a word, may produce a noticeable click. However, because the deficit retards the timing and hence prevents further large deficits, this occurs only infrequently and is not irritating. At the end of the trace the network delay is decreasing and the reserve is correspondingly growing. If this state of affairs persists the mechanism shown in Fig. 8.3 comes into play.

The behaviour of the protocol discussed above is based on examples from calls connected through ring–ring bridges. Connections made around a single ring behaved in a manner nearer the ideal, as would be expected. Calls made across the satellite link, whilst producing quite good voice quality for most of the time, suffered from two problems not found on local connections. The first problem was that the overall delay was considerably larger than necessary, that is in comparison with inevitable signal propagation delay caused by the use of the satellite. The second was that under certain conditions of network loading, successive voice packets would be delivered each with a slightly longer delay than the previous one. This could occur for quite a long series of samples and, because of the way the play-back mechanism operated such a series of deficits led to a series of clicks in the earpiece. There was insufficient time during the project to find the cause of this latter problem.

The problem of the larger than necessary delays on satellite calls was due to a number of factors:

(a) the large TDMA frame size used in the satellite bridge;

(b) the lack of any priority or reservation mechanism for the transmission of voice packets by the satellite bridge; and

(c) the necessity to use rather long packets for voice samples, due to lack of processing power in the Z80s controlling the phones.

As explained in Chapter 4, one of the main considerations in the design of the high-performance satellite bridge was the production of hardware sufficiently powerful to allow short TDMA frame size and to be able to cope with priority and reservation schemes.

Two human factors issues were noted during the course of the experiments, both related to silence suppression. The first concerns the clipping of the ends of words. This comes about because the energy level in the sound of words, particularly words ending in vowel sounds, tends to die away and may fall below the silence threshold before the word has ended. The simple solution of making the software continue to transmit for a set period, typically 0.5 to 1.0 s, after the end of the last non-silent burst produced quite acceptable sound quality. The beginnings of words were not found to suffer this problem.

The other issue is associated with the difference perceived by the listener between background noise, when this is actually transmitted across the network, and the silence that prevails when no blocks are transmitted. The difference between background noise and complete silence was found to be quite noticeable, particularly when a caller in a quiet room listens to words spoken in a room with a high ambient sound level. Most listeners found this rather unpleasant. A simple solution that was tried was to lengthen the 'silence' period transmitted after a talkspurt to 2 s and then, when blocks failed to arrive, to play back repeatedly the last second of sound received, assuming this to represent background noise at the transmitting end. This worked well most of the time, but on occasion the so-called silence contained some distinct feature that could be clearly heard recurring at 1 Hz. The relative success of this simple method suggests that it should be possible to synthesize a sensible background noise substitute. With a processor more powerful than the Z80, it would be possible to collect together a number of intervals of 'silence' taken from the ends of talkspurts and to use these in combination to avoid the repetition problem.

8.2.5 *Review*

Voice activities in the project were carried out on quite a small scale, but nevertheless gave some encouraging results in one area, that of producing a viable protocol for dealing with the variable delay in packet delivery. This protocol, in conjunction with the LVC service offered by the network, showed promising potential for handling voice in the packet environment. It had the additional advantage of easily allowing the incorporation of silence transmission suppression. To establish packet voice as a technology suitable for use in an operational network, a number of additional problems must be satisfactorily resolved. In a network carrying voice, or more generally any data with delay constraints, there must be some method of blocking new traffic from the network, in order to protect existing traffic. Having allowed the initiation of a telephone conversation, it is plainly quite unacceptable to allow subsequent traffic growth to interfere with this conversation. In circuit-switched networks, this blocking of new calls occurs automatically; when all available circuits are in use, no new calls can be initiated since there are no free circuits to take these calls. This applies even in the case where circuits only exist on a statistical basis, as, for example, when a TASI scheme is employed. As has been discussed in Chapter 3 (see Section 3.7.1), a similar effect might be achieved in packet networks by enabling the network to restrict the number of virtual calls requiring guaranteed bandwidth and transporting the packets belonging to these calls with priority. These ideas were not tested in the project.

Clearly the requirement to carry a mixture of voice and data has substantial performance implications for all the components of a packet network. In addition to the normal packet load from data traffic, such a network would have to handle with priority large numbers of small packets and incorporate some form of bandwidth accounting to underpin call blocking. No attempt was made in the design of the equipment used in the project to take these requirements into account; such an attempt would have taken the project completely outside its timescale. The new design of satellite access controller described in Chapter 4 does attempt to deal with the requirements of voice traffic. This controller is designed to cope with a high packet rate. It also has built-in features to assist with the fragmentation of data packets necessary to allow use of the short TDMA time frame required for voice. All the other network equipment would require similar redesign.

8.3 Image transmission

The ability to handle image data is of fundamental importance in information technology. Images, particularly high-quality colour images, are a source of very large quantities of data and their handling and transmission present a major challege to the capabilities of any network. It was thus highly desirable to demonstrate that such data could be handled on the UNIVERSE network in a manner having potential to provide useful services. The video analogue of the telephone is the image station. The station transmits images captured from a video camera across the network and displays images received from the network on a video monitor. As with voice, images may be transmitted between stations in a real-time fashion or transmitted from a station to an image storage system for subsequent retrieval. However, the analogy between voice and video activities must not be taken too far. The 'still' video image, that is a single video frame which is displayed continuously, is an important class of image that has no corresponding voice analogue and must be handled in its own right.

During the project a number of image stations were built by Loughborough and located at several sites on the network, including the demonstration site. The design of these stations is described below. They were used to handle still images and to provide a 'slow-scan' system, that is a system providing a series of views of a scene or object that are updated automatically every few seconds. These are examples of video services that might be accommodated in a network operating in the transmission speed range of the UNIVERSE network.

8.3.1 *Television standards*

To ensure that an image station can interwork with commonly available devices, such as cameras, display monitors, and video recorders, it is essential that the design conforms to relevant television standards. These include specifications for line, field, and frame timing, and for the amplitude and bandwidth of the video and synchronization signals. The UK standard employs 625 lines per frame, with the frame refreshed 25 times per second. To help reduce flicker, an interlaced method of scanning the frame is used. Each frame is considered to be made up of two fields, a field containing all the odd-numbered lines and a field containing the even-numbered lines. A whole frame is scanned by scanning the lines of the first field followed by the lines of the second. Thus a video signal consists of equal-length periods of visual signals, which generate the lines of the image, separated by inter-line gaps, this pattern being periodically broken by longer inter-field and inter-frame gaps. In the 625-line system, the line period is $64\,\mu s$ (i.e. $1/(25 \times 625)\,s$) which is split as $52\,\mu s$ for the visual line signal and $12\,\mu s$ for the interline gap. The bandwidth of the signal may extend to 5 MHz and thus the sampling rate for good resolution should be of the order of 10 MHz. This correponds to the use of 520 picture samples per line or, more conveniently, 512 samples per line. The inter-field gaps consist of 25 blank lines giving 575 lines as the total number of lines in the frame that contain visual information. With 512 samples per line, the maximum sampling rate is 10^7 samples per second and the average rate is $512 \times 575 \times 25$ or 7.36×10^6 samples per second.

The amplitude of a monochrome signal may be digitized in a linear fashion as the signal amplitudes are defined to lie within fixed limits. Using eight bits per sample provides extremely good resolution and, in fact, six bits per sample produces pictures of very reasonable quality. Thus for black and white digital television a bit rate of the order of 50 Mbit/s is required.

To produce colour television pictures, three video signals, one each for the red, green, and blue components of the visual signal, are necessary. For television broadcasting these are combined in an analogue fashion into a single composite colour video signal. There are three different schemes, called NTSC, PAL, and SECAM, that are widely used for carrying out this combination. In the UK, the PAL (phase alteration line) scheme is used. Since colour information is perceived by the eye with lower resolution than luminance information, it is possible to use a lower bandwidth for this composite signal than three times that used for black and white pictures. In fact, the PAL signal is contained within a bandwidth of 8 MHz. Correspondingly it is possible to use 16 bits per pixel, rather than 24, to code the colour amplitudes (see below). Thus, for colour digital television a bit rate of the order of 120 Mbit/s is required.

Bit rates of this order are very difficult to handle and, given the transmission capability available to the project, reductions of at least two orders of magnitude are called for. In general there are two ways of reducing the data rate; by removing some of the redundancy in the signal and by lowering the quality of the signal. Usually there are strong correlations between the colour values of adjacent pixels within a frame and these can be used to reduce the data required to represent the frame. An even stronger correlation will also usually exist between frames and in fact parts of the picture may remain unchanged over many frames. Frame updating, that is only transmitting changes to the current frame to produce the next frame, is a very effective way of reducing the data rate. Image quality may be reduced by using fewer pixels per line, fewer lines per frame, fewer frames per second, and by using fewer bits to code the amplitude at each pixel. These techniques must be implemented in combination to produce the required large reductions in bit rate.

8.3.2 *The image stations*

The main component of an image station is a framestore: this is a memory device which insulates the rest of the station from the rigid timing regime of conventional video cameras and monitors. This store must be able both to capture a frame from the standard video signal output of a camera and to refresh a conventional monitor at the normal frame rate. To carry out these functions, a framestore must contain digital to analogue, analogue to digital, and control circuitry, as well as memory. The remaining parts of an image station are a controlling processor and interfacing equipment for the network.

Loughborough based the design of the image station on use of the Multibus standard. This choice was influenced by the availability of several components, including

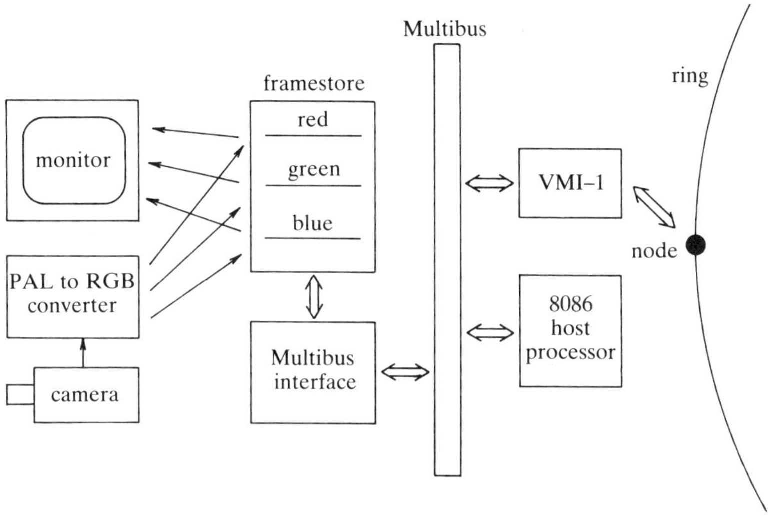

FIG. 8.5 Image station configuration.

- the Logica Polynet VMI-1 Cambridge ring interface, which could sustain data rates up to 800 kbit/s and included an on-board implementation of the basic block protocol;
- single-board computers (including on-board memory), based on the Intel 8086 processor; and
- software development facilities at Loughborough for the Intel 8086 processor.

A framestore board, developed by BTRL for slow-scan video applications, was used to build the framestore. This board design contained a complete subsystem including all the necessary circuitry for picture acquisition and display, together with memory and a microprocessor interface. Three of these boards were used to give a full colour display; one of the boards provided the basic timing signals for synchronization and the other two were driven as slaves from this. The arrangement of the hardware of the image station is shown in Fig. 8.5.

To allow good colour coding while at the same time limiting the representation of each pixel to 16 bits, the red, green, and blue signals were transformed to luminance and chrominance signals (Y, U, V) between the camera and the framestore and in reverse between the framestore and the monitor. Bit allocations were six bits to the luminance Y and five bits each to the colour difference signals U and V. (Y, U, V) signals are linearly

TABLE 8.2 Resolution options

pixel/line	*line/frame*	*interlace*	*number of frames*
512	512	yes	1
512	256	no	2
256	256	no	4

related to (R, G, B) signals; this relationship takes the following form for the PAL television standard:

$$\begin{aligned} Y &= 0.299R + 0.587G + 0.114B, \\ U &= -0.147R - 0.289G + 0.437B, \\ V &= 0.615R - 0.515G - 0.100B. \end{aligned}$$

Use of this coding produces good-quality colour pictures from 16 bits per pixel. Monochrome images could be handled easily by making use of the luminance component on its own.

Each image station could capture and display images from sources conforming to the UK 625-line standard. Each frame was sampled during the active line periods and the resulting pixel data stored in the framestore memory. Three spatial resolutions were available, as listed in Table 8.2.

The three framestore boards were interfaced to the Multibus by a specially-built interface board that mapped the I/O registers on the framestore boards onto Multibus addresses. Given the design of the framestore board, this was the best access arrangement possible. However, it limited access to the framestore memory to the low rate of roughly 16 Kpixel/s. On the other side of the processor, the VMI-1 acted as a full bus master transferring data and control information by DMA, thus allowing much higher data rates.

As well as the access speed limitation imposed by its register-driven interface, the BTRL framestore board was of rather inflexible design, requiring straps for changes of resolution and a change of components to change the interfacing arrangement. Accordingly, Loughborough embarked on the design of a framestore to work with Multibus, having fast access to the frame data and being fully configurable under processor control. A prototype of this new store was available by the end of the project but was too late to be used for applications experiments.

8.3.3 *Image system organization*

The image handling system was organized in a fully distributed fashion around three types of components: image stations, control computers providing the user interface, and picture storage systems. The BBC microcomputer was used for system control, providing the user with a menu-driven procedure for selecting services from the system repertoire and providing appropriate parameters. These computers controlled the activities of the other components, transmitting commands as SSP requests. Both large server systems and GEC 4000 systems were programmed to act as picture stores. Image data was transferred between image stations and between an image station and picture store over UNIVERSE LVCs.

In the later stages of the project, equipment was built to provide remote control of the video cameras for demonstration purposes: this is described in Chapter 10. The complete repertoire of commands available to control a remote image station was as follows:

- Capture a single frame from an external video source.
- Display a specified frame from those already stored within the image station. (*Note:* Depending on the resolution, one, two, or four frames are available in an image station (cf. Table 8.2).)
- Fetch a single frame from another station.
- Start and stop fetching a continuous sequence of frames from another image station.
- Control of initialization and monitoring.
- Pan, tilt, or zoom the camera.

Clearly, not all of these commands are appropriate to a picture store. After selecting one of these services, the user is prompted for the name of the target station and any associated parameters. The control station uses an SSP request to convey the request to the target station, expecting an acceptance or rejection reply which is displayed to the user. If the service requested requires the target station to fetch an image or images from another station, it attempts to open a circuit to this source station. If this is successful, the fetching station makes simple requests for images on a frame by frame basis. To ensure continuity of data when a continuous stream of images is requested, an extra request is sent at the outset so that the transmitting station always has an outstanding request at the end of each frame transmission. The image data is transmitted line by line with line addresses; frames are separated by a frame header.

Flow control, which is not provided by the network, is effected by the simple strategy of arranging that the sender is always slower than the receiver. The receiver is provided with sufficient buffering capacity to accommodate any bunching of packets that may occur within the network. In the event of lost or corrupted data, a recovery mechanism is provided. For single frame transfers a tally is kept of lines correctly received and, at the end of the frame, any lost lines are requested again. Continuous transfers do not use this feature, as it is likely that the affected area of the image will be correctly updated during the next frame.

A scheme for data compression using an interframe coding technique known as 'block conditional replenishment' was implemented. Here, the frame to be transmitted is segmented into small blocks having a similar aspect ratio to the whole frame. Taking each block in turn, the luminance amplitude of each pixel is compared with that of the corresponding pixel in the block when last transmitted and the modulus of the difference calculated. The peak and average differences are found for the block and compared with perception thresholds. The block is transmitted if either threshold is exceeded. In addition to this conditional transmission, all blocks are transmitted unconditionally at a slow rate. This ensures that the whole picture is updated periodically.

8.3.4 *Performance*

The maximum data rate into and out of the image station was limited by the framestore hardware to approximately 260 kbit/s. A limit of 220 kbit/s was artificially imposed on transmission from a station to ensure that the receiving station could always keep up. This lower rate could also be managed without problem by the satellite bridge. The stations were normally operated with a resolution of 512 pixels per line and 256 lines per frame. Packing of lines was two per full-size basic block. At a rate of 220 kbit/s, which corresponds to 13.4 full-size blocks transmitted per second, this allows a transmission rate of approximately 27 lines per second or one frame every 9 seconds for uncompressed data. To evaluate the performance of the system, it was operated for long periods both on a local ring and between two sites across the satellite network. In both cases there were periods of exclusive use as well as shared use.

As might be expected, operated in continuous replenishment mode on a local ring, the system showed no sign of being subject to interference from other activities on the ring. The system also worked well across the satellite in conditions where there was little other traffic flowing through the satellite bridge. However, with substantial additional traffic, data started to be lost before the satellite bridge reached its capacity limit, the limit being on the receive side (see Chapter 4). Overcoming these limitations in the performance of the satellite bridge was one of the main concerns in the redesign of the bridge, which is discussed in Chapter 4.

Work on compression, using the block conditional replenishment algorithm, was only completed after the demise of the OTS satellite and could only be tested between stations on a local ring. Even with the very slow rate of frame acquisition, substantial reductions in data rate could be achieved with quite acceptable results. Unfortunately using the 8086 processor as the compression engine, the computing task of compression took a little longer than the time to transmit an uncompressed image!

There are a number of subjective comments that can be made about the image handling system. Notwithstanding the severe limitations on performance, the system looked interesting. It was possible to transfer, store, and retrieve good-quality colour images with ease in quite reasonable time. Even with the very slow refresh rate in continuous transmission mode, it was possible to gain a good impression of activities at a remote location. In this case the ability to control the camera, as was done for the demonstrations (Chapter 10), from the remote end was very important. The distributed arrangement of the system was generally found to be convenient and flexible. As has been described, the system allowed several people, usually at different locations, to control the functioning of all the equipment. No constraints on users' actions were implemented and this would clearly be necessary in an operational system. It was obvious that the problem of orchestrating the control activities of even two users is an interesting challenge.

8.4 Image data compression

As has been mentioned in the previous section, the digitization of video images, particularly sequences of moving images, generates very large quantities of data. As can be seen from the parameters described in Section 8.3.1, the raw data rate produced by digitizing the video signal from a camera is more than 1000 times that of a speech signal. Consequently, image data compression has been the subject of considerable research effort over a number of years, particularly since large, low-cost memories and single-chip, high-speed digital signal processors have become available. It was not in any way part of the brief of the project to carry out research into methods of compression, rather the project was in the position of being a potential customer for these research efforts. As such a customer, Logica undertook for the project a survey of methods for both still and moving images and carried out evaluations, both theoretically and by simulation, of some promising possibilities. This section gives an outline of the results of these studies.

The subject of image coding for compression purposes is described in detail in the book by Clarke (1985) and in the two review articles by Jain (1981) and Netravali and Limb (1980). The reader interested in pursuing this subject is referred to these works which contain extensive bibliographies.

8.4.1 *Characterization of single images*

Evaluations of image compression algorithms are concerned with two factors, the compression ratio achieved, that is the number of bits in the input image divided by the number of bits in the compressed image, and the quality of the reconstructed image. Some algorithms produce a fixed compression ratio irrespective of the input image, in which case the quality of the reconstructed image is a function of the input image. Other algorithms achieve a fixed quality for the reconstructed image with the compression ratio being a function of the input image. In this class are the 'reversible' algorithms, algorithms that allow the original image to be reconstructed exactly. For highly 'random' images these algorithms may produce compression ratios less than unity. Most algorithms fall somewhere between these two extremes. In all cases, however, the overall performance of any algorithm, that is the combination of compression ratio and image quality, is a function of the particular input image. To evaluate algorithms it is, therefore, necessary to define properties that can serve to classify images to enable comparisons to be made and to define measures that assess the quality of the reconstructed image.

For single monochrome images, the following statistical quantities have been found to be useful.

(a) *Variance, V* This is a measure of the level of contrast in the image. It is defined as

$$V = \left\langle (I_{xy} - \langle I_{xy} \rangle)^2 \right\rangle ,$$

where I_{xy} denotes intensity at pixel coordinates (x, y).

(b) *Entropy, E* Image entropy gives some measure of the information content of an image. It is defined as

$$E = -\sum_{I=1}^{2^B} P(I) \log_2 P(I),$$

where $P(I)$ is the probability of a given pixel having intensity I and B is the number of bits used to digitize the pixel intensities. The concept of signal entropy comes from information theory and it can be shown that, if the intensities at different pixels are all independent of each other, the minimum number of bits that can be used to code each sample in any compression scheme which allows perfect reconstruction is E bits/sample. In the absence of any correlation information about an image, $P(I)$ is simply equal to $n(I)/N$, where $n(I)$ is the number of pixels having intensity I and N is the total number of pixels. E has a maximum value when all intensities are equally probable. In this case $P(I) = 1/2^B$ for all I, from which it follows that $E = B$ bits.

(c) *Correlation,* C_{ab} The correlation between the intensities of pixels from two sets, a and b, of image data can be used as a measure of the extent to which the intensity at any given pixel is related to the intensities at neighbouring pixels. If a is the set of pixel intensities of the original image, then correlations with set b, obtained either by shifting one pixel in x or y or by shifting in a diagonal direction, are of interest. The correlation is given by

$$C_{ab} = \rho_{ab}/\sqrt{V_a V_b},$$

where ρ_{ab} is the covariance of the two sets of data defined by

$$\rho_{ab} = \langle (I_a - \langle I_a \rangle)(I_b - \langle I_b \rangle) \rangle .$$

A colour video picture is formed by the superposition of red, green, and blue images. Each of these images can be characterized by the quantities described above. In general, there is a high correlation between the intensities at the same pixel in each of the three images. Scenes can be quite well recognised from any one of the three images displayed on its own. This correlation may be summarized by a colour covariance matrix given by

$$\rho = \begin{pmatrix} \rho_{\mathrm{rr}} & \rho_{\mathrm{rg}} & \rho_{\mathrm{rb}} \\ \rho_{\mathrm{gr}} & \rho_{\mathrm{gg}} & \rho_{\mathrm{gb}} \\ \rho_{\mathrm{br}} & \rho_{\mathrm{bg}} & \rho_{\mathrm{bb}} \end{pmatrix} \equiv \begin{pmatrix} V_{\mathrm{r}} & \rho_{\mathrm{rg}} & \rho_{\mathrm{rb}} \\ \rho_{\mathrm{gr}} & V_{\mathrm{g}} & \rho_{\mathrm{gb}} \\ \rho_{\mathrm{br}} & \rho_{\mathrm{bg}} & V_{\mathrm{b}} \end{pmatrix},$$

where ρ_{rg}, ρ_{rb}, and ρ_{gb} are the covariances of the red and green, red and blue, and green and blue sets of intensities, respectively. The matrix is, of course, symmetric. A corresponding colour correlation matrix C may also be defined. In this case the diagonal elements are all unity.

The only measure used in the study to assess the quality of the reconstructed images was the root mean square (RMS) error, defined as

$$\sigma_{\mathrm{E}} = \sqrt{\left\langle (I_1 - I_2)^2 \right\rangle},$$

where I_1 and I_2 are the intensities at corresponding points in the original and reconstructed images respectively. Although this is a useful measure in the evaluation of image quality, the important arbiter of quality is the human eye. Thus a compression algorithm which produces quite acceptable σ_{E} may introduce subjectively annoying features that make the reconstructed image unacceptable.

8.4.2 *Compression algorithms*

Image compression algorithms may be divided into three broad classes.

(a) *Coding schemes* These are the schemes that are used to code the intensity information at each pixel location. Included are both methods used to reduce the redundancy in the colour information and schemes to use variable-length code words, where frequently occurring intensity levels are given short code words, whilst rarely occurring levels have long code words.

(b) *Predictive methods* Here the intensity at a given pixel is compared with a predicted value based on values at neighbouring pixels, or, for changing images, values in previous frames. The difference in the two values can in general be transmitted using fewer bits than are required for the absolute value, as the distribution of difference values is usually sharply peaked about zero.

(c) *Spatial transformation methods* Two-dimensional unitary transformations may be applied to the data, yielding sets of uncorrelated coefficients which can be efficiently quantized and coded.

To produce large compression, schemes from these different classes must be combined.

The colour redundancy in a given image can be removed by finding a linear transformation of the colour intensities (I_r, I_g, I_b) that diagonalizes the colour covariance matrix. This type of transformation is called a Karhunen–Loève (KL) transform. It has the property that, whilst the diagonal elements of the original covariance matrix will be roughly equal, the corresponding elements of the transformed matrix are strongly ordered in magnitude,

$$V_1 \gg V_2 \gg V_3.$$

It is thus possible to use less bits to code the values of the second and third transformed variables. Derivation of these results can be found in Clarke (1985) and Jayant and Noll (1984).

The KL transformation outlined above is the basis for the transformation to (Y, U, V) colour intensity coordinates used in analogue TV broadcasting and mentioned in Section 8.3.2. The transformation from (I_r, I_g, I_b) coordinates to (Y, U, V) coordinates, given in that section, is for most scenes quite close to the ideal KL transformation, with Y the dominant coordinate. In the transformed coordinates, Y, the so-called luminance, carries most of the scene variance and U and V carry the colour information. Since the latter coordinates carry only a small proportion of the scene variance

TABLE 8.3 Example of Huffman coding

Intensity level	*Probability*	*Code*
4	0.26	10
5	0.24	00
3	0.20	01
6	0.15	110
2	0.05	11100
1	0.04	11101
7	0.03	11110
0	0.03	11111

and the eye is relatively insensitive to small colour changes, they can be heavily compressed.

Coding may also be used to compress data by allocating variable-length codes to intensities, short code words being assigned to intensity levels that occur with high probability. Huffman coding (see, for example, Hamming, 1980) provides optimal results for this type of encoding. An example of Huffman coding for three-bit intensity levels is shown in Table 8.3. Given a monochrome image with N pixels and intensities distributed with the probabilities shown in the table, the number of bits required to encode the image, using the Huffman encoding shown in the third column, would be

$$2 \times (0.7N) + 3 \times (0.15N) + 5 \times (0.15N) = 2.6N \text{ bits},$$

compared with the $3N$ bits that would be required with straightforward encoding. Bearing in mind that the code itself must also be transmitted, the gain is very small. This is not surprising since inter-pixel correlation has not been taken into account. In general, used on their own, encoding schemes of this type produce only small compressions of the data. Their importance comes when used in conjunction with the 'prediction' and 'transform' methods described below.

Predictive methods are based on the idea of exploiting the redundancy in images by predicting the intensity at a given pixel from intensities that have been transmitted previously. Since this same prediction can be made both when the image is compressed and when it is reconstructed, only the difference from this prediction need be transmitted. Compression is achieved because the variance of this intensity difference is normally smaller than the variance of the absolute intensities. Prediction functions are of the form

$$I^* = \sum_i w_i I_i,$$

where I^* is the predicted intensity and w_i are weights subject to

$$\sum_i w_i = 1.$$

The summation may include previous pixels in the same line, pixels in previous lines or in previous frames. Differences, $\epsilon_{i,j}$, are transmitted, where

$$\epsilon_{i,j} = I_{i,j} - I^*_{i,j}.$$

Clearly some initial absolute values $I_{i,j}$ must also be transmitted. The following functions were used in the study of single image compression.

(a) *Pixel difference*

$$I^*_{i,j} = I_{i-1,j}.$$

The predicted intensity is the intensity of the previous pixel and the $\epsilon_{i,j}$ which are transmitted are the pixel differences.

(b) *Linear extrapolation along line*

$$I^*_{i,j} = 2I_{i-1,j} - I_{i-2,j}.$$

(c) *Parabolic extrapolation along line*

$$I^*_{i,j} = 3I_{i-1,j} - 3I_{i-2,j} + I_{i-3,j}.$$

(d) *Pixel average*

$$I^*_{i,j} = (I_{i-1,j} + I_{i,j-1})/2.$$

The predicted intensity is the average of the previous pixel and the corresponding pixel in the previous line.

For image sequences the functions investigated were the following.

(e) *Frame difference*

$$I^*_{i,j,k} = I_{i,j,k-1};$$

k is the frame number and the $\epsilon_{i,j}$ which are transmitted are differences in intensity of corresponding points in successive frames.

(f) *Pixel difference of frame difference*

$$I^*_{i,j,k} = I_{i,j,k-1} + I_{i,j-1,k} - I_{i,j-1,k-1}.$$

TABLE 8.4 Rice–Plaunt coding scheme

Difference	*Code word*
0	1
1	01
−1	001
2	0001
−2	00001
⋮	⋮

(a) *Coding of differences*

Triplet	*Code word*
000	0
001	100
010	101
100	110
011	11100
110	11101
111	11111

(b) *Huffman encoding*

Having found the differences, it would be possible to code these using Huffman coding. In practice, a computationally simpler method, due to Rice and Plaunt (1981), produces compressions that approach those of Huffman codes. Since this method is fixed for all images, there is no need to transmit the code. In the method, each difference value is initially coded using the scheme shown in Table 8.4(a). The coded differences are then blocked into groups of n values; the corresponding code will then contain exactly n ones. If the length of the code block is greater than $3n$, zeros will predominate. In this case the sequence is further compressed by breaking it into groups of three bits and using the Huffman code shown in Table 8.4(b). If ones predominate and the length of the block is less than $3n/2$, the same coding is applied to the complement of the block code. For intermediate cases, no further coding is applied to the block.

This method of coding allows error-free reconstruction of the image. In general, some number of the least significant bits of the frame differences will be randomly distributed. If these are removed by shifting down before coding with the Rice–Plaunt scheme and then restored on reconstruction by shifting up and adding an offset, very good approximations to the original image can be obtained with much larger compression.

In 'spatial transformation' methods, the image is first divided into square blocks of $N \times N$ pixels, with say $N = 4$ or 8, which are then transformed by a two-dimensional unitary transformation. The transformations used are generally separable and symmetric, in which case the transformation can be written

$$C_{pq} = \sum_{x=1}^{N} \sum_{y=1}^{N} I_{xy} k_{xp} k_{yq},$$

where C_{pq} is the transformed image block. Since the number of transformed 'intensities' C_{pq} is equal to the number of intensities I_{xy} in the block, this does not, at first sight, seem to result in any gain. However, transformations are chosen so that the C_{pq} are approximately uncorrelated, and thus the redundancy associated with the original intensities, I_{xy}, is removed. This normally results in only a few of the C_{pq} having large values and the others having values distributed around zero with small variance. The latter may be ignored or grossly quantized using a small number of bits.

Two transformations were studied: the discrete cosine transformation, in which the image is represented by cosine functions of varying frequency, and the Hadamard transformation, which is a similar transformation with the cosine replaced by a square wave function. In particular, a variant of the Hadamard transformation developed by Rice (1975) was implemented for evaluation.

In general, transform methods require considerably larger computational effort than predictive methods but can be used to produce higher compression ratios.

8.4.3 *Performance of algorithms*

The algorithms described in the previous section were evaluated both theoretically and empirically to determine the range of compression factors to be expected. Theoretical analysis of algorithms seeks to give estimates for the compressions to be expected from given algorithms as functions of the statistical properties of the input images and the allowed distortion in the reconstructed image. However, since the simple statistics that can be used in theoretical studies are far removed from full characterizations of real images, these studies must be complemented by extensive empirical investigations using real images. In general, the theoretical estimates proved to be quite good, although, as might be expected, almost always gave optimistic results for compression ratios. They may be regarded as realistic estimates of the upper limits for the compressions which may be obtained. Since the evaluations undertaken of the algorithms described above were detailed and lengthy, only a brief summary of the empirical results is given; these, however, serve to give some idea of the possibilities.

For single image tests, a set of five images was digitized. This set consisted of an image of the surface of Mars (MARS), a LANDSAT image of Death Valley in California (DTHVLY), two 'head and shoulder' images (referred to as FACE and FIZOG), and an image of the surface of the Moon (MOON). A resolution of 512×512 was used, with the luminance intensity sampled at eight bits. The variance, entropy, and correlation statistics for these images are shown in Table 8.5. It will be seen that the image of the Moon is the most active, having high variance and entropy and low correlations. The face images are both highly correlated.

TABLE 8.5 Variance, entropy, and correlations of test images

Image	*Variance*	*Entropy*	*Correlations*		
			X	*Y*	*Diagonal*
MARS	92.2	4.88	0.925	0.971	0.914
DTHVLY	663.0	6.51	0.950	0.965	0.932
FACE	966.2	6.69	0.995	0.993	0.990
FIZOG	122.6	7.00	0.997	0.996	0.995
MOON	2324.0	7.14	0.928	0.960	0.922

Huffman coding, the predictive method with Rice encoding, and the Hadamard transform method of Rice were programmed in FORTRAN and applied to the images. For the prediction method it was found that using the average of the intensities from previous pixel and the corresponding pixel in the previous line gave the best results, being slightly better than simple pixel difference encoding. The results for reversible compression schemes are shown in Table 8.6. The compression figures shown in brackets are for full Huffman coding of the differences.

TABLE 8.6 Compressions obtained with various schemes

Image	*Huffman*	*Pixel difference*	*Pixel average*	*Rice transform*
MARS	1.62	2.87	3.1	2.82
DTHVLY	1.24	1.55 (1.72)	1.60 (1.90)	1.47
FACE	1.19	1.96 (2.31)	2.07 (2.46)	1.80
FIZOG	1.13	1.97 (2.31)	2.08 (2.47)	1.92
MOON	1.11	1.26	1.42	1.18

Using the 'pixel average' prediction formula, the compression tests were repeated in a non-reversible mode, that is with truncation of least significant bits, to obtain reconstructed images that were virtually indistinguishable from the initial images. Table 8.7 shows that the compression was roughly doubled with RMS errors in the range 1.2 to 1.4. The Rice transform method was also repeated for FACE and DTHVLY with various truncations. The best compression ratios obtained, again with undiscernable changes from the initial images, are shown. The RMS errors for these transformations were about 3.

The tests carried out were based on the goal of achieving compression with little or no distortion in the reconstructed image. These tests showed that the predictive method using pixel averages performed well and for low

TABLE 8.7 Compressions with truncation

Image	*Pixel average*	*Rice transform*
MARS	6.39	
DTHVLY	3.23	4.53
FACE	4.60	6.28
FIZOG	4.61	
MOON	2.25	

distortions produced similar compressions to those obtained with the chosen transform method. The latter, however, involves considerably larger computation. For compression with higher distortion but producing visually acceptable results, transform methods can be used to produce higher compression ratios.

Some performance measurements were also made on the compression of moving sequences of images. One sequence of frames appropriate to video conferencing was used. This consisted of 100 frames, captured at 25 frame/s and digitized with 512×512 pixels at eight-bit precision grey levels, that is only the luminance intensity was digitized. The sequence showed a moving head and shoulders against a static background. Unfortunately, because of lack of suitable equipment it was not possible to play back the reconstructed sequence at full frame rate.

The reconstructed frames were visually acceptable after averaging pixels over 2×2 blocks and reducing the sampling accuracy to six bits. These measures reduce the initial rate of 50 Mbit/s to 9.375 Mbit/s. Two methods of reducing this bit-rate were investigated, coding frame differences, that is taking the prediction for a pixel to be the intensity at the corresponding pixel in the previous frame, and lowering the frame sampling rate. Clearly these methods interact since the correlation between frames depends on the frame interval.

Sampling the input image sequence only at intervals of n frames gives a compression ratio of n; the display of the reconstructed image sequence must nevertheless be refreshed at the full frame rate to avoid flicker. The simplest scheme is to display each image n times and accept the increased jerkiness of the restored sequence. This jerkiness can be reduced by interpolating intermediate frames or by the technique of transmitting a different $1/n$th of the pixels from each of the frames of the n-frame cycle. If the pixels of each fraction of the frame transmitted are distributed uniformly over the frame in a random manner, jerkiness is minimized. On reconstruction, $1/n$th of the pixels are updated in each display refresh, giving a visual impression of a continuously changing image. This scheme is known as dot interlace.

It was predicted that the reconstructed sequence quality would be acceptable at a sampling rate of one frame in four. Using this rate, frame differences and pixel differences of frame differences were coded using the Rice method. Frame differences were also used with a dot interlace scheme and with frame averaging. In frame averaging, the average of two frames is used to reduce the frame rate by a factor 2 and then alternate averaged frames are transmitted to give a factor 4 reduction in frame rate. All these methods produced approximately similar results. The factor 4 reduces the bit rate to 2.34 Mbit/s and the difference encoding reduces this by a further factor of between 3 and 3.5 to produce a final bit rate of between 0.7 and 0.8 Mbit/s, a rate within the capabilities of a network of the UNIVERSE type.

8.5 Conclusion

In this chapter, the work carried out in the project on the transport of two media, digital voice and video, has been described. It will have been seen that the construction of a communications service for a single media involves research and development in a variety of technologies. Protocols must be developed that take into account the characteristics of the underlying network, methods of digitization and compression schemes must be evaluated, and the resulting design implemented in suitable hardware. The same basic service may be presented in a number of forms: in the traditional dialogue form; in a messaging form using intermediate storage; and in a multi-cast form, linking several users of the service. A good design will encompass all of these forms. With a more complex service, such as that of linking image stations, the problem of the organization of control becomes serious. Who is allowed to do what, and when? Considerable ingenuity, backed by a great deal of experiment, is required to produce control regimes that provide flexibility, prevent anarchy, and yet are simple and natural to use.

To produce a network offering multi-media services, further major steps are required. Firstly, the transmission network must be able to carry the resulting traffic without interference between services; for example, voice data must not be disturbed by large file transfers and transactions must not be delayed by video traffic. Secondly, there must be suitable means for organizing the services for different media to work together to provide an integrated multi-media service. Offices may be equipped with means for voice communication, with means for video and facsimile communication, and with devices for accessing computing services. The problem is to give the user of this office the ability to orchestrate the use of this equipment to allow communication and collaboration with colleagues in similar remote offices. Although no serious work on this problem was undertaken in the project, the potential difficulties were clear to see in the project demonstrations (Chapter 10) which showed several services working side by side. Clearly, all aspects of multi-media communications services will be fruitful research areas for many years yet.

9. Other experiments

In the previous two chapters, experiments characterized as being either distributed in nature or concerned with visual and aural communication have been described. Within the project, experimental work in a number of other areas was pursued, which either made direct use of the facilities provided by the UNIVERSE network or could be regarded as complementary investigations necessary for any more realistic exploitation of the networking principles explored by the project. The first three, full-screen terminal access, videotex, and teletex, may be regarded as applications in the area of special terminal services. The topic of remote file access, described next, was another look at a familiar networking problem, that of devising a file transfer scheme that is simple to implement on a variety of computers. Encryption, fault-tolerant rings, and high-speed bulk data transfer, concern matters fundamental to network communications. It should be pointed out that this latter work was carried out during the later stages of the project and had no influence on the design of the UNIVERSE network. GEC had an especial interest in the token ring and constructed one as part of the work on fault-tolerant rings. This was attached to the network, where its presence had the fortuitous side-effect of serving as a check that features specific to the Cambridge ring LAN technology did not become embedded in the UNIVERSE network design.

9.1 Full-screen terminal access

Remote login to time-sharing systems has traditionally been one of the main service activities associated with computer networks. Using simple terminals and line editors, remote users have been able to make very effective use of the time-sharing services offered by mainframe and minicomputers. In the mid-1970s, these time-sharing systems started to offer local users full-screen terminal access, particularly for editing, making use of more advanced terminals. Access in this form is hard to provide for remote users since both high data rates in bursts and short transmission delays are required to give good performance to the user. Since the UNIVERSE network had the possibility of coping with the former, there was clearly interest in determining whether problems arising from the long transmission delay could be overcome to give remote users a satisfactory full-screen service. To be generally useful, this investigation had to be undertaken within the constraint that existing host software remained unchanged.

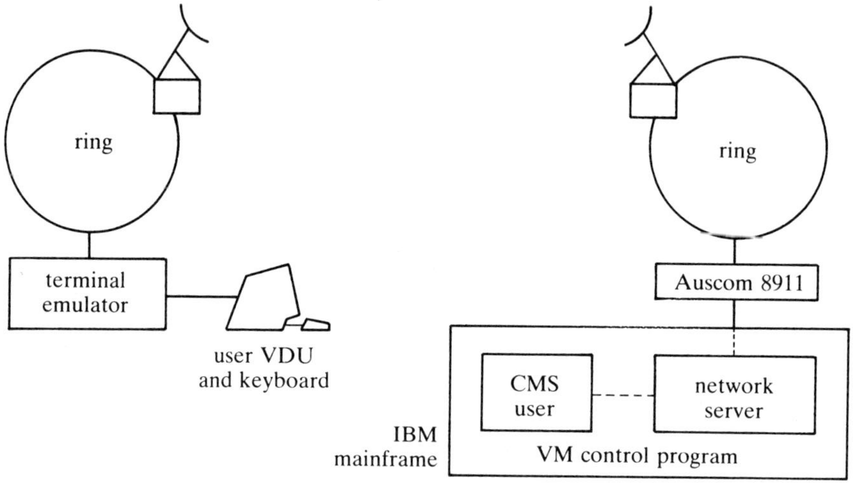

FIG. 9.1 **Full-screen access components.**

The particular example chosen for experimentation in the project was full-screen access to the IBM time-sharing operating system VM/CMS, as this was available at RAL and was interfaced to the Cambridge ring (Chapter 6). Access is normally provided by locally attached 3270-type terminals (IBM, 1980 and 1987). In the experiment, the activities of a remote 3270 terminal were emulated by a dedicated microprocessor system driving an ordinary visual display unit and keyboard terminal (VDU). Standard communication between a 3270 and an IBM computer is carried out by the exchange of buffers. In the experiment, these buffers were carried, essentially unaltered, as basic blocks on a UNIVERSE LVC. The aim of the implementation was to minimize the user's perception of the delays associated with the transmission of these buffers.

9.1.1 *Implementation*

Emulation of the 3270 terminal was implemented on a dedicated microprocessor driving a standard ASCII VDU via an RS232 link. The VDUs used (Cifer models 2605 and 2634 and the BBC microcomputer running a terminal emulator) would all accept commands to move the cursor and perform various control functions. The emulation was programmed in BCPL running under the TRIPOS operating system for both an Intel 8086 processor and for the 68000-based large server. The IBM mainframe computer was interfaced to a Cambridge ring via an Auscom 8911 channel interface (Chapter 6). A network server program to handle communication with remote terminals was written for the IBM machine. The program ran in its own virtual machine, handled the interface to the Cambridge ring and implemented the basic block and LVC set-up protocols. Within the VM environment, programs running in a virtual machine appear to communicate directly with terminals. This communication, however, is intercepted by the VM control program, which actually handles real terminals. A feature of the control program is that this communication can be diverted to another virtual machine, which can substitute for the expected real terminal (IBM, 1979). This feature was used to link virtual machines to the network server program and thence to their remote terminals. The general arrangement is shown in Fig. 9.1. The terminal emulator and the network server program were linked via an LVC and exchanged data buffers carried in basic blocks. A simple transaction protocol with an acknowledgement window of one was used. The network server program dealt with data transparently, only handling OPEN and OPENACK. It was able to handle simultaneously a number of remote terminals linked to different virtual machines.

The display is changed by both the incoming data stream from the IBM host and by user keyboard actions. The emulator program keeps a buffer of 1920 bytes holding a copy of the current 24 lines of 80 columns displayed on the screen. The screen may be split into fields, each having an attribute character in the first position describing the characteristics of the field (protected, highlighted, etc). Buffers of data arriving from the host contain orders that process the contents of the screen buffer and which may be dependent on the field structure of the image. The emulator program decides which parts of the display are changed by new data and sends appropriate streams of characters to the VDU to effect these changes on the screen. Since the connection between the terminal and emulator microprocessor was quite slow (9.6 kbit/s), some optimizations were made in the emulator to minimize the flow of data to the screen.

9.1.2 *Acceptability*

This activity was given rather low priority within the project and the implementation was only completed late in the life of OTS. Consequently, no detailed evaluation was carried out but some limited use of remote access was made. It was clear that the main factors affecting the perceived quality of the terminal are:

(a) the response time for simple commands;

(b) the response time when the display is being changed rapidly, as, for example, when a user is browsing through a file; and

(c) the allocation of keyboard functions.

Some simple keyboard actions can be dealt with locally by the emulator but, for simple commands that require access to the host, the response time is largely governed by the network round-trip delay. For communication across the satellite network this delay is about one second, which is long, but in many circumstances is not disturbingly long. It is also fixed and regular, which is a great deal more acceptable than unpredictable delay. There is no way of overcoming this limitation, short of a radical alteration in the roles of host and terminal to ensure that the host plays no part in any simple operation.

The speed at which the whole display can be changed, in an operation such as browsing through a file, is governed by a number of factors including the network delay, the network bandwidth, and the bandwidth of the connection between the emulator and the terminal. Some measurements were made of the time taken to skip through substantial files. These gave times a factor 2.5 to 3 longer than equivalent times using a terminal locally connected at 1 Mbit/s. This represents a worst-case situation in which the user always takes very little time to recognize that the next page is required. Since the network bandwidth is high, the reduced speed of display is caused by the network delay and the limited bandwidth from the emulator to the terminal. The latter could be removed by using a personal computer with integral screen handling as the remote terminal. This would reduce the factor between the browse times for local and remote terminals to less than 2.

It was possible to find a reasonably satisfactory mapping between the keys of the particular terminals used and those of a real 3270 terminal and, although the subjective acceptability of a terminal used in conjunction with a particular system is difficult to quantify, this was not thought to be a fundamental weakness.

The general acceptability of a terminal is a very subjective matter, highly dependent on the expectations of the user and the details of the tasks being undertaken. Generally, it was felt that the emulation was quite successful, being more than acceptable when viewed in relation to other remote access possibilities but falling a good deal short of the performance of a locally connected terminal. Although not done in the project, it would have been interesting to have built a terminal emulation into a personal computer with an integral, memory-mapped display, as it was clear that this would have removed a number of weaknesses.

9.2 Videotex

Videotex is a computer-based information retrieval system characterized by its use of inexpensive terminals based on television receivers. There are two forms of the system, interactive videotex and broadcast videotex. In both forms, the unit of information is a complete screen image or page. Associated with videotex is a character code which, as well as representing the usual alphanumeric characters, also defines codes for a set of mosaic characters from which simple geometric shapes may be constructed. This code also defines shifts that allow the use of eight colours, including black and white. Broadcast videotex, known as teletext, involves the sending of pages of information within television broadcast transmissions. It is a non-interactive service in the sense that the user has no return channel with which to manipulate the contents of the information base. Subscribers to an interactive videotex service normally access the database service via the public switched telephone network. British Telecom has operated a public interactive videotex service, called Prestel, in the UK for a number of years. Similar public services have been constructed in other countries and some private services exist within institutions. It was only this interactive form of the service that was of interest within the project and below it will be referred to simply as videotex.

A feature of large videotex systems is the distribution of the information base across a number of server computers situated at different locations. The contents of the information base is organized on a regional basis: information of local interest is stored in the appropriate regional database; information of general interest is replicated in each of the regional databases.

BTRL and UCL were interested in implementing a videotex service on the UNIVERSE network in order to gain some experience with such a service in a packet-switched, LAN-based environment, including exploring possible protocols. They also wished to investigate possible enhancements to the service enabled by the use of such a network. The enhancements envisaged were:

(a) inclusion of video images in the range of information that could be held in the database;

(b) co-ordination of access to the distributed servers so that the whole system would appear to the user as a homogeneous information base which could be searched without reference to the location of individual pages;

(c) use of broadcast protocols to distribute updates to the database servers; and

(d) improvements in the management of the system.

In the event, only the inclusion of image data (a) was pursued beyond the planning stage. The other enhancements were planned around the introduction into the system of a new component, the *videotex service agent.*

This agent had features in common with both the *filing machine* (Section 7.3) and the *file distribution agent* (Section 7.4). In the new arrangement, each client would be associated with a 'local' videotex agent. This agent would retrieve pages from local and remote servers in response to the client's commands and would set up connections to remote servers as required. It would use some reliable broadcast scheme to convey page updates to all the servers in the system. The agent would also undertake a number of management tasks, such as authentication of clients and the collection of accounting and monitoring information.

Implementation of such a system is a substantial piece of work and it was decided that insufficient effort was available to produce worthwhile results before the end of the life of OTS. However, a full videotex service was implemented on the network with enhancements to allow video images to be stored and accessed. The implementation, which is described below, was used extensively at BTRL, at two of the project demonstrations (Chapter 10), and to a limited extent at UCL.

9.2.1 *Implementation and performance*

BTRL based the development of a videotex database server for the UNIVERSE network on the hardware and software used for the Prestel service. GEC 4000 machines are used for this service and hardware and software to interface these machines to the Cambridge ring was available (Chapter 6). This reduced the development work to the manageable, if still quite substantial task of giving the Prestel software a new service interface for access from the Cambridge ring. Prestel is a connection-oriented service and LVCs were used for sessions between client terminals and the server.

Three types of videotex terminal were implemented, a standard alpha-mosaic terminal, a photo-videotex terminal, and an advanced photo-videotex terminal. A BBC microcomputer was programmed to act as a standard videotex terminal, supporting the full alpha-mosaic character set. On request by a user for a videotex session, the terminal set up an LVC to the videotex server, in the usual fashion. Thereafter, exchange of basic blocks was used throughout the session, with the server echoing requests and then sending the retrieved data. The maximum size of an alpha-mosaic videotex page is 1 Kbyte and so a page could always be sent in a single basic block. No recovery from network errors was provided by this experimental system (though basic blocks with an invalid checksum would generally be discarded); the user noticing anything amiss with the retrieval of a page was expected to repeat the request. As part of other research activities, BTRL had also developed two experimental photo-videotex terminals. In addition to alpha-mosaic information, the first of these terminals could display a full-colour video image insert of a size up to one-sixth of the screen area.

It had been designed to operate at 4.8 kbit/s over a standard V.24 interface, using Hadamard transform coding for compression to give page retrieval times in the region of 15 s. It was a straightforward matter to link this terminal, unchanged, to a BBC microcomputer running a slightly modified version of the software described above, to produce a photo-videotex terminal attached to the network. The second terminal was aimed at a videotex service making use of a 64 kbit/s integrated services digital network (ISDN) channel and was capable of displaying a high-quality full-screen image in less than 10 s. This terminal was built around the Multibus standard and included an Intel 8086 single-board computer. It was possible to interface this to the Cambridge ring using a Polynet VMI-1 board but in this case a fair amount of software modification was required.

For photo-videotex frames, the insert or full-frame picture data are stored and transmitted as a set of additional 1 Kbyte pages linked to a standard, alpha-mosaic page. Picture inserts consist of up to 8 Kbyte of additional data; full-frame pictures need up to 64 Kbyte. The handling of image data raises some new considerations. Since image reconstruction from compressed data is a relatively slow process, either the terminal must be able to buffer a whole picture in compressed form or the server must control its rate of transmission of blocks to suit the speed of the reconstruction process. Also, for image data transmission, errors are more serious since compression schemes typically introduce structure and dependencies within the data. Loss or corruption of data in a single block may render complete image reconstruction impossible. Thus some form of automatic correction procedure becomes essential, although none was implemented in the project.

From the point of view of response, the most stringent testing of the videotex implementation was in the use of the simple alpha-mosaic terminal. The design specification for the Prestel service sets a maximum response time of 2 s and it was clearly important to meet this objective. Conditions in which the network was most heavily loaded prevailed during the demonstrations at the Telecom '83 exhibition at Geneva. Under the heaviest conditions of ring loading, retrieval times from a local server were in the range 2–3 s, and times for retrieval across the satellite network, again heavily loaded, were the expected second or so longer.

9.3 Teletex

Teletex was defined by the CCITT in 1980 as an international service for the automatic memory-to-memory transfer of text documents between word-processors. The service could be carried on any type of network but three public networks receive particular mention in the specification, namely, the switched telephone network and both packet-switched and circuit-switched data networks. These are to be used to link teletex terminals together to form the teletex service network, in much the same way that the telex network is constructed. International gateways to allow communication between terminals attached to different types of network are defined, as is a gateway to the existing telex service. The general requirements for the teletex service are set out in CCITT recommendation F.200 (CCITT, 1981*d*); specifications for teletex terminals and the protocols to be used for the service are detailed in recommendations S.60, S.61, S.62, and S.70 (CCITT, 1981*e*). The issue of these recommendations roughly coincided with the start of the project and it was thought at the time that teletex was likely to become an important new telecommunications service. In fact this has not come about and, although the service is quite extensively used in a few countries, it has not gained widespread acceptance on an international scale.

Three of the participants, BTRL, Logica, and UCL, were interested in gaining experience in the teletex protocols and their implementation and decided to pursue this within the project. The intention was to build teletex facilities at each site using quite different implementation strategies, to achieve interworking between these facilities across the UNIVERSE network, and to make some evaluation of the results. Details of these implementations and the results of the tests are described below.

The general specifications for a teletex terminal, which are set out in recommendation S.70, are quite comprehensive. The recommended data rate for the terminal is 2.4 kbit/s and use of the terminal in local mode for document preparation, display, and printing should not be interrupted by communications activities. The page is the basic unit for text transmission and presentation, and in S.61 a large character repertoire is specified, which includes, in addition to the definition of over 300 graphic symbols, a wide range of formatting functions for controlling the layout of printed pages. Recommendation S.62 defines a session layer protocol for the service. The recommendations are completed by S.70, which defines a transport service layer based on the assumption that it will be supported by an X.25-like network service.

For communication across the UNIVERSE network it was decided that the CR82 protocol (Section 6.5) would be used as the network service for S.70. Although this made for some duplication between the functions provided by S.70 and those already present in the lower layers, this was considered to be simple to implement.

9.3.1 *Implementation*

At BTRL, a teletex workstation was built using Multibus-based, board-level microprocessor components. The familiar Polynet VMI-1 provided the hardware interface to the ring. As well as keyboard and display, the station was equipped with floppy disks and a printer and could handle the full teletex character set. The software ran under the control of Intel's real-time multitasking operating system, RMX-86, which allowed all the functions provided by the terminal to be operated concurrently.

By contrast, the Logica implementation was purely a software package, designed to run in a time-sharing environment and to be highly portable. This software, which was called Teleport, was written in C. Portability was achieved by defining system-independent interfaces to the presentation and network layers and by separating the support software into system-dependent and -independent components. For the UNIVERSE implementation, interface software to the CR82 network service was written and system dependent components were provided to enable the package to run under Unix.

At UCL, the main objective was to integrate the teletex service with other existing telecommunications services, rather than produce a self-contained terminal. To achieve this, a distributed systems approach was employed for the implementation, using a number of systems and services already available on the local network. Only the S.70 transport layer and the S.62 session layer were new implementations. Document creation and presentation were carried out via the word-processing tools available on one of the Unix systems, thus limiting the character set to ASCII.

9.3.2 *Results*

Experimental work concentrated on establishing full interworking among the three implementations. Testing was carried out in three stages. Since the systems ran on different hardware and under different operating systems, three different implementations of the CR82 network service were needed. Compatibility between the latter was established by simple text file transfers between pairs of sites. In the second stage, the individual teletex implementations were tested looped-back to themselves; this was possible as the teletex specification requires terminals to be able to receive and transmit concurrently. Finally, interworking tests were carried out, aided in the first instance by the physical portability of the BTRL terminal which was taken to both UCL and Logica.

As might be expected, the interworking tests revealed a number of faults in each of the implementations. Considering the complexity of the standard, the range of features offered, and the general lack of experience with the specification, rather few problems were encountered. Most problems stemmed from two causes, different interpretation of the wording of the standard and failure to implement the full range of allowed formats for parameters specified or implied in the standard. After these faults had been diagnosed and remedied, exchange of documents in each direction between the implementations was possible.

Conformance rather than performance was the goal for this work and figures for throughput rates on complete documents were disappointingly low, being of the order of 100 characters per second. This is less than half the recommended rate for the public service and also much lower than any limitation attributable to the use of CR82 on a satellite network. Tests of the separate implementations using a looped-back, dummy network with no delays, showed that they were rather slow but probably just fast enough for the standard. Knowledge gained of the properties of the teletex protocols suggested possible optimizations of the software but these were not applied. Obviously, an alternative to CR82 would be necessary for any serious use of a network with satellite transmission segments.

The goal of producing implementations of the teletex protocols tested by interworking was achieved. Unfortunately, no critical assessment of the teletex protocols was made either from the point of view of implementation or performance. Nevertheless, the industrial partners were able to gain substantial experience on which to base commercial work in this field and in fact Logica were able to make some direct commercial exploitation of the Teleport software.

9.4 Remote file access

To meet the requirements of distributed computer systems operating over a high-speed wide area network, GEC decided to investigate the problems of defining a protocol which would combine the properties of a traditional file transfer protocol with provision of a remote form of interface to the facilities typically provided by the filing system component of an operating system. The aims for this protocol were that it should be suitable for use on a wide range of computing systems, interfacing to their existing filing systems and utilities, that it should be easy to implement and have performance characteristics appropriate to the UNIVERSE network. The work was influenced by two existing file transfer mechanisms: the UK interim standard file transfer protocol (NIFTP) (DCPU, 1981*a*) used on JANET and the fileserver protocol (Dion, 1981) used in the Cambridge distributed system.

9.4.1 *The protocol*

The remote file access (RFA) protocol provides facilities for an application process to access and manage files held on a remote filing system. The protocol operates over a reliable transport service which is assumed to deal with error recovery and flow control and which preserves sequentiality. A record structure is imposed on the data stream, with a record length field preceding each record. Records may be of two types, control or data; data also may be of two types, text or binary. For text files, each record represents a line. A number of access functions are defined for files: create and delete; read, write, and append; print; and change file parameters (e.g., protection or access). In addition, two access functions are defined for directories: create and list.

To initiate a request for access, the client sends a 'request' control record to the filing system (server). This record contains two mandatory parameters in an agreed format: the access function and the data type (text or binary). Any additional parameters required are presented as text strings to be interpreted by the filing system, which should provide defaults where necessary. There is no negotiation of parameters and only sequential access to files is supported.

After a successful request, data records may be transferred between the client and the filing system. Records that are longer than the maximum allowed in the protocol may be sent as 'split' records for reassembly at the destination. An 'end-of-file' control record is also defined. When an access is complete, the filing system sends a 'successful access' control record. On receipt of this the client may either start a new request for access or disconnect the transport service connection. If any protocol errors are detected by either end during a request, a transport service 'reset' message is sent. Non-fatal filing system errors (e.g., space exceeded) are indicated to the client by a 'warning' control record.

An optional facility in RFA was implemented for authentication, based on the Cambridge mechanism described in Chapter 7. The principles were the same as those used for authenticated BSP (Section 7.5.3): the client process passes a representation of itself (or its user) as an RFA parameter to the filing system. This is checked in the usual way and if it is acceptable, the access is allowed to proceed. A mapping service associated with the filing system may also be consulted, by presenting the representation, to obtain details of the access restrictions and directory context appropriate to the client.

9.4.2 *Implementation*

Implementations of the RFA protocol were written for three quite different types of system. These were Intel MDS systems running the single-user ISIS-II operating system, GEC 4000 systems running the multi-user OS4000 operating system, and PDP 11 systems running under Xenix (a Unix-based operating system). All three implementations operated over CR82 (Section 6.5) and the GEC 4000 version was also capable of using the OS4000 JANET transport service interface to X.25. The file structures of both the ISIS-II and Xenix systems are based on data streams with no record structure. In both these implementations, arbitrary fixed-length record boundaries were imposed on binary data. In text mode, each character was analysed to find line terminators, to enable the restructuring of the data into line records.

In addition to performance testing, the RFA protocol was used operationally in the USE experiment (Chapter 7) and also routinely between machines at GEC.

9.4.3 *Results*

The protocol was successfully implemented on three widely differing machines, and there were few problems in establishing compatibility. All the implementations were interfaced to the existing filing systems. In some instances special utilities had to be written to permit RFA to carry out management tasks; however, in practice a unified interface to local and remote files could be provided to the user. The protocol could have been extended relatively easily to encompass random access to files. This was not done partly because it was thought that, given the high bandwidth and long delay of the network, operations such as editing a file could be more efficiently achieved by first transferring the file to local storage, updating it, and then transferring it back. Whichever method is used, locking mechanisms are required for this type of operation on a shared file store and these were not provided. The protocol was not as easy to implement as had been hoped and in some cases rather extensive knowledge of the filing system was required to implement some of the RFA functions. For non-record-based file stores, converting text to record format is CPU-intensive. While structuring data into records in the protocol solves the problem of distinguishing control information from data, use of this same record structure to represent significant structure within the data leads to well-known problems (instanced in this case by the appearance of the 'split' record mechanism).

Performance measurements were made of the protocol running under a variety of circumstances. Using CR82 for memory-to-memory transfers between a pair of GEC 4065s attached to a Cambridge ring, data rates between 128 kbit/s with 64-byte records and 205 kbit/s with 1000-byte records were obtained. For disk-to-disk file transfers between the same two systems on the ring, a rate of 48 kbit/s was achieved using RFA. Over the satellite the rate using RFA was only 20 kbit/s but this was a reflection of the use of the CR82 transport protocol with its window of one rather than a measure of the performance of the protocol. In a direct comparison with an implementation of NIFTP, over an X.25 link running at 19.2 kbit/s the transfer data rate was between 15 and 16.8 kbit/s, depending on record size, much faster than the 5.5 kbit/s measured for NIFTP.

In conclusion, one may remark the similarity between the goals of RFA and recent standardization work resulting in the definition of ISO file transfer, access, and management protocol (ISO, 1987). Both are concerned to combine into a single protocol the functionality of file transfer, file access, and a variety of file system management activities. Differences in the concrete realizations of file stores, such as those encountered here in implementing RFA, have led to considerable effort in standardization work being directed towards the formulation of a suitable abstraction of a file store in terms of which such protocols may be defined.

9.5 Encryption

In many commercial applications of computer networks it is essential that some data transmissions are protected against unauthorized interception or modification. Because of the broadcast nature of the down-link in satellite communications, networks using satellite transmissions are particularly vulnerable to interception of data. As part of the project, Logica and GEC undertook the design and implementation of a system of secure end-to-end communication across the UNIVERSE network. The scheme, which has since become quite familiar, was based on using the Data Encryption Standard (DES) algorithm, published by the National Bureau of Standards of the USA in 1977 (NBS, 1977; Beker and Piper, 1982), to encrypt data and on using the Rivest–Shamir–Adleman (RSA) (Rivest *et al.*, 1978) form of public key cryptography (PKC) to distribute securely the keys associated with the DES algorithm. Encrypting and decrypting were carried out in the host processors on user data. No work was carried out on the problems associated with encrypting header information from the lower levels of protocol or of developing encryption servers.

9.5.1 *Implementation of the data encryption standard*

The DES algorithm employs a 56-bit key, which both sender and receiver must know, and enciphers data in 64-bit blocks. After an initial permutation, the data is subjected to 16 identical iterations of a function which combines substitution and transposition under the control of a 48-bit key. In each stage of iteration, a different 48-bit key is used, the set of 16 keys being derived from the initial 56-bit key by iterated permutation, shifting, and selection of bits. The output from the final iteration undergoes a 32-bit swap and the inverse permutation of the initial permutation to produce a 64-bit output block of cipher text. Deciphering is accomplished using the same algorithm with the set of 16 keys used in reverse order. A discussion of the algorithm and its vulnerability to attack can be found in Denning (1982); simplified descriptions of the algorithm may be found in Davies (1979) and Tanenbaum (1981).

The DES scheme may be used in a simple fashion in which each 64-bit data block is enciphered independently from all others. This is suitable for many applications but does suffer the weakness that identical input blocks produce repeated cipher text blocks. This deficiency may be eliminated by employing a mode of the DES called cipher block chaining, which utilizes well-known properties of the boolean 'exclusive-or' (XOR) operation applied bit-wise to the 64-bit data blocks. Each enciphered text block is XORed with the next data block; the result is encrypted to form the next cipher text block. A 64-bit initialization vector is required, in addition to the key, to begin the process. Deciphering is achieved by decrypting a block and then XORing it with the previous encrypted block. Chaining in this way removes repeated blocks of cipher text. It also protects against a type of cryptanalytic attack known as 'chosen-plaintext' attack, where the cryptanalyst is able to choose plaintext messages, cause these to be enciphered, and examine the resulting ciphertext. Use of DES in both basic and cipher block chaining modes was implemented in the encryption experiment.

GEC made a hardware implementation of the DES algorithm on a Multibus board using an Am9518/AmZ8068 data ciphering processor chip to carry out the encryption. The board also contained hardware for direct memory access, allowing encryption of blocks of data to be carried out in a single operation; it was used for tests in an MDS development system. A software implementation for the DES was written in C by Logica for a PDP 11/44 minicomputer running under the Xenix operating system. Tests between these were made using CR82 (Section 6.5) as an end-to-end transport protocol.

9.5.2 *Encryption using public keys*

To be able to exchange encrypted messages using the DES algorithm, both the sender and the receiver must know the key being used. Since it is generally necessary to change keys regularly whatever the encryption scheme used, exchanging keys is a significant problem. Use of a public key cryptography (PKC) scheme for the key exchanges largely eliminates this problem. As with the DES, PKC schemes use a standard, publicly known function to encrypt and decrypt the message. However, for PKC schemes there are two keys, an encryption key and a decryption key. Only the receiver knows the decryption key and the scheme is arranged so that it is impractical to calculate the decryption key from a knowledge of that for encryption. Therefore, the encryption key can be made public and the problem of exchanging keys secretly is avoided. However, since the decryption key is unknown to the sender, a transmitter who mislays the plain text cannot retrieve it from the enciphered text!

PKC schemes depend on the use of 'one-way' operations, that is, operations where it is easy to produce a result but where knowing the result does not give any reasonable method of discovering the starting conditions. For example, starting from two large prime numbers p and q, it is easy to produce a large composite number $m = p \times q$ but, by choosing sufficiently large p and q, it can be made impractically difficult to find p and q from a knowledge of m. The RSA algorithm depends on the impracticability of factorizing numbers having only very large factors. For the RSA algorithm, as in the DES, the message is split into blocks of digits, but for RSA these blocks are treated as unsigned integers. Encryption is carried out by raising M_i, the integer representing the ith message block, to a certain power s and then keeping only the remainder modulo m. Thus, the ith encrypted block, E_i, is given by

$$E_i \equiv M_i^s \pmod{m}. \tag{9.1}$$

The numbers s and m form the public key for the receiver to whom the message is being sent. The encrypting prescription (9.1) is not secret. Decrypting is achieved by the inverse operation

$$M_i \equiv E_i^t \pmod{m}, \tag{9.2}$$

where t is the private key known only to the receiver. Now provided s and $\phi(m)$ are co-prime, that is the greatest common divisor (gcd) of s and $\phi(m)$ is 1, s and t are related by

$$st \equiv 1 \pmod{\phi(m)}, \tag{9.3}$$

where $\phi(m)$ is the Euler function of m; moreover the solution for t is unique. $\phi(m)$ is defined as the number of positive integers less than m which are co-prime with m, that is the number of integers k such that $1 \leq k < m$ and $\gcd(k, m) = 1$. Derivation of equation (9.3) and its solution for t are discussed below. The important point for the RSA scheme is the evaluation of $\phi(m)$.

If p is a prime number, then each of the numbers $k = 1, 2, 3, \ldots, p-1$ is co-prime to p and therefore $\phi(p) = p - 1$. Thus if the value of m were chosen to be prime, $\phi(m)$ would be trivial to evaluate and calculation of the private key t would be as easy for the public as for the proposer of the public key. However, if m is chosen to be the product of two large primes p and q, that is $m = p \times q$, then $\phi(m) = (p-1)(q-1)$ and can be calculated from a knowledge of p and q, whereas starting only from a knowledge of m, calculation of $\phi(m)$ is equivalent to factorizing m, which can be made a prohibitively large calculation for large m.

The RSA algorithm was implemented in C, using prime numbers p and q of 256 bits. Prime numbers of this magnitude can be generated by randomly choosing a 256-bit odd number and testing it for primality. If the number is not prime, it is incremented by two until a prime is found. For a 256-bit number, a mean of approximately 130 tests are required to find a prime. Using the Rabin method (Rabin, 1976) for primality testing and a PDP 11/44, a pair of primes, p and q, and the associated t (see above) could be produced in an average time of a little over two hours. In the project, the exponent s was always taken as the prime number $2^{16} + 1 = 65\,537$.

The RSA scheme can also be used as an authenticating mechanism. A sender A, having encrypted a message using the intended receiver B's public key, applies a second stage of encryption using his own private key. The receiver B then applies a first stage of decryption using A's public key and then a second stage using his own private key. If the decrypted text contains A's *signature* (already known to B), B can be sure that A is the originator of the message and that the message has been transferred in secrecy.

The RSA encryption scheme makes use of a number of results from number theory. These results can be found in any standard work on the subject, such as Hardy and Wright (1979) or Burton (1980). Readers interested in a less formal account of the mathematics involved are referred to Devlin (1984). Here, only very brief indications of the derivations are given.

If equation (9.3) holds, that is

$$st \equiv 1 \pmod{\phi(m)},$$

then for some integer k

$$st = k\phi(m) + 1. \tag{9.4}$$

Using (9.1), calculating modulo m, it follows that

$$E_i^t \equiv M_i^{1+k\phi(m)}.$$

Euler's theorem, a well-known result from number theory, states that

$$M^{\phi(m)} \equiv 1 \pmod{m},$$

for all M co-prime with m. Thus

$$E_i^t \equiv M_i \pmod{m}.$$

Note that for the method to work, the messages M_i must be co-prime with m, but the chance of this not being the case is extremely small.

Solving equation (9.3) for t can be achieved by noting that the equivalent form, (9.4), is a linear Diophantine equation in two variables, t and k, i.e.

$$st - k\phi(m) = 1,$$

where s and $\phi(m)$ are known numbers. Such equations can be solved by using the Euclidean algorithm to find the gcd of the coefficients s and $\phi(m)$, which is known to be one from the choice of s and m, and then working back through the algorithm to find t and k. Finally, the value of $\phi(pq)$ can be found by noting that, if p and q are prime, of the $(pq-1)$ positive integers less than pq, only the $(p-1)$ multiples of q and the $(q-1)$ multiples of p amongst these are not co-prime to pq. Thus

$$\phi(pq) = (p-1)(q-1),$$

a particular case of the general result that

$$\phi(pq) = \phi(p)\phi(q),$$

for p and q co-prime.

9.5.3 *Use of the encryption scheme*

The combination of the RSA and DES algorithms was used to provide a solution to the key management problem. For good security, frequent key changing is required and to reduce the overhead involved in the calculation of PKC keys, a tiered arrangement of keys was used. To produce a sequence of DES keys, it is only necessary to generate a series of independent random 56-bit numbers. This is an easy task and consequently the following three-tiered scheme was used.

(a) *PKC keys.* These were generated locally by each user and the public part distributed to all other users. Since the generation process is slow, these keys are only changed infrequently.

(b) *Master keys.* These are DES keys which are generated by users and exchanged between pairs of participants using RSA encryption. These keys are changed periodically.

(c) *Session keys.* These are DES keys which are generated for each session and which are exchanged using DES encryption with master keys.

Using this hierarchical arrangement for key management and the hardware implementation of the DES, encrypted data could be transferred across the network with only a small percentage addition in time compared with non-encrypted data.

9.6 Fault-tolerant rings

At first sight it would appear, from the point of view of reliability, that the ring topology has serious disadvantages as the communications arrangement for a LAN. A ring is formed by a series of point-to-point links joined by active repeater nodes and a single failure of a repeater, loss of power to a repeater, or a break in a transmission cable will cause the network to fail completely. In a well-known paper, Saltzer *et al.* (1981) have compared ring and passive bus LAN techniques and concluded that neither has a clear advantage over the other. In particular, rings have the property that equipment faults should be easy to locate. This feature is exploited in ring designs that use wiring centres to give easy maintenance and repair. Aside from any theoretical analysis, it is well established that soundly engineered rings perform with a degree of reliability that is quite acceptable for most applications.

Within the project, it was recognized that in some circumstances the basic lack of resilience of the ring would not be acceptable and both GEC and LUT carried out work aimed at providing fault tolerance for rings. For their part, GEC developed a fault-tolerant ring based on the token-passing strategy for sharing the bandwidth of the LAN. In their scheme, the ring is always configured from all working links and repeaters and no reliance is placed on stand-by equipment for dealing with faults. LUT efforts were directed towards slotted rings. Since for slotted rings particularly, the maximum point-to-point transfer rate falls with increases in the length of the ring (see Chapter 2), any wiring centre scheme has disadvantages when applied to the Cambridge ring. The LUT work was directed at obtaining the advantages of the wiring centre without sacrificing performance.

For neither of the two fault-tolerant schemes described here was work taken beyond the stage of testing basic ideas. To move beyond this stage would require very careful designs for chip implementations of the components, together with detailed theoretical analysis to quantify the level of fault tolerance to be expected.

9.6.1 *Background*

No attempt at fault-tolerance is built into the design of the Cambridge ring. Instead, various precautions have been taken to reduce the likelihood of repeater failure; in particular, the part of the repeater required to keep the ring in operation has been made as simple as possible and this part of the repeater is powered from the ring rather than via the attached station. In practice, these precautions have produced a system that is highly reliable and which is quite suitable for situations where rare failures are not catastrophic.

The simplest method of allowing for repeater failure is to provide each repeater with a relay, which is actively held on by the repeater to bring it into the ring. If the repeater fails, the active signal is expected to be removed, which will cause the ring to bypass the repeater. This simple scheme has the serious disadvantage that, as repeaters fail, arbitrarily long lengths of link without repeaters can appear in the ring. One solution to this problem is to separate the relays from the repeaters by placing the relays in a wiring centre.

In the MIT design suggested by Saltzer and Pogran (1980), the relays are housed in a single unit, the wiring centre, creating a ring in the shape of a star with the data path travelling in both directions along each spoke to a repeater and station. A further line in each spoke is used to hold on the corresponding relay. Since the wiring centre is modular, unpowered and purely passive, it is unlikely to fail completely. This approach does have the disadvantage that great lengths of cable may be required. The IBM architecture (Andrews and Schultz, 1982) allows multiple wiring centres in the ring to overcome this problem. However, this leaves the ring vulnerable to failures in the links between wiring centres and these must be protected by a different method.

A 'self-healing' technique has been known for many years (Zafiropulo, 1974) and is used in a number of ring designs, for example the Racal Planet ring (Carrington, 1983; Böschen, 1982), as well as to provide protection between wiring centres in the IBM design. This technique uses spare links passing all the way round the ring in the opposite direction to the normal links. Repeaters can detect breaks in the ring on their upstream side. If one is detected, the repeater 'loops back' by taking its input from the spare link entering it from the opposite direction. This repeater must also signal to the repeater immediately upstream from the break to 'loop back' in the opposite sense, that is to connect its output to the spare link. Thus, after a failure the ring becomes 'linear' in form with the signal passing in both directions through each repeater node.

A number of other schemes for providing fault-tolerance have been proposed, including various 'braided' structures in which extra links are provided to bypass intervening repeaters. A comparison of these techniques can be found in Falconer (1984).

9.6.2 *A reconfiguring fault-tolerant token ring*

The basic idea of the GEC scheme for a fault-tolerant ring is that the network consists of a mesh of repeaters and links that is configured into a ring on startup or after a fault has been detected. All links between repeaters are bidirectional, so that the data passes in both directions along each link. Each repeater is capable of attaching a station and may have one, two, or three links. The repeaters are joined together by an arbitrary mesh of links. A repeater with a single port can only be used to connect a station on a spur attached to the mesh by another repeater. A repeater with two ports can be inserted in any link and used either to attach a station or to extend the length of the link. It is the repeaters with three ports that determine the loop topology of the network and these are called 'intelligent' repeaters. A distributed reconfiguration procedure carried out in these 'intelligent' repeaters configures the mesh of bidirectional links into a ring which passes in both directions along each operational link.

The tolerance of this network in the presence of link or repeater failure is due to the presence of loops in the mesh. Using this property, such a network can be arranged to be tolerant of a large class of failures. If no stations are connected via spurs but are instead all inserted in loops in the topology, single faults and many multiple faults will not partition the network. Although the GEC work was concerned with the development of a token-passing ring based on this type of configuration, the basic ideas are largely independent of the access method used on the ring and could be applied to most ring-based LAN technologies.

The 'reconfiguration' procedure takes place when the network is powered up, after a network failure is detected, or when new links or repeaters are added into the network. The basis of this procedure is that it is initiated by one repeater which sends out a startup signal that is passed around the network until it is received back at the initiating repeater, having traversed each usable link in both directions. The ring is then formed from the path traced by this startup signal. The links of a repeater are arranged in a defined cyclic order; when the signal arrives at a repeater for the first time, or arrives back at it on a link by which it left it earlier, it is directed to leave by the next available link in this cyclic order. If, however, the signal arrives at a repeater that it has already visited, along a link on which it did not previously leave, it is looped back along the same link by which it just arrived. It is not difficult to see that this procedure produces a ring in which each sound link is traversed in both directions.

In the procedure described above, all 'intelligent' repeaters are capable of being the initiating repeater and the initiating repeater has no special role once reconfiguration is complete. To complete the procedure two practical problems must be solved: methods must be found for identifying working links and for deciding which repeater is to initiate any reconfiguration. To determine the set of working links, the startup signal is immediately preceded by a probing signal. This probing signal is sent first on any link and is returned at once by the repeater at the end of the link. If the probing signal is not received back quickly, the link is bypassed and not included in the ring. Reconfiguration of the network may be requested by any station attached to an intelligent repeater. This initiation will normally occur after a ring-broken state has been detected by a repeater and notified to a station. To prevent multiple reconfigurations being started after failure or power-up, each station waits for a time dependent on its address before

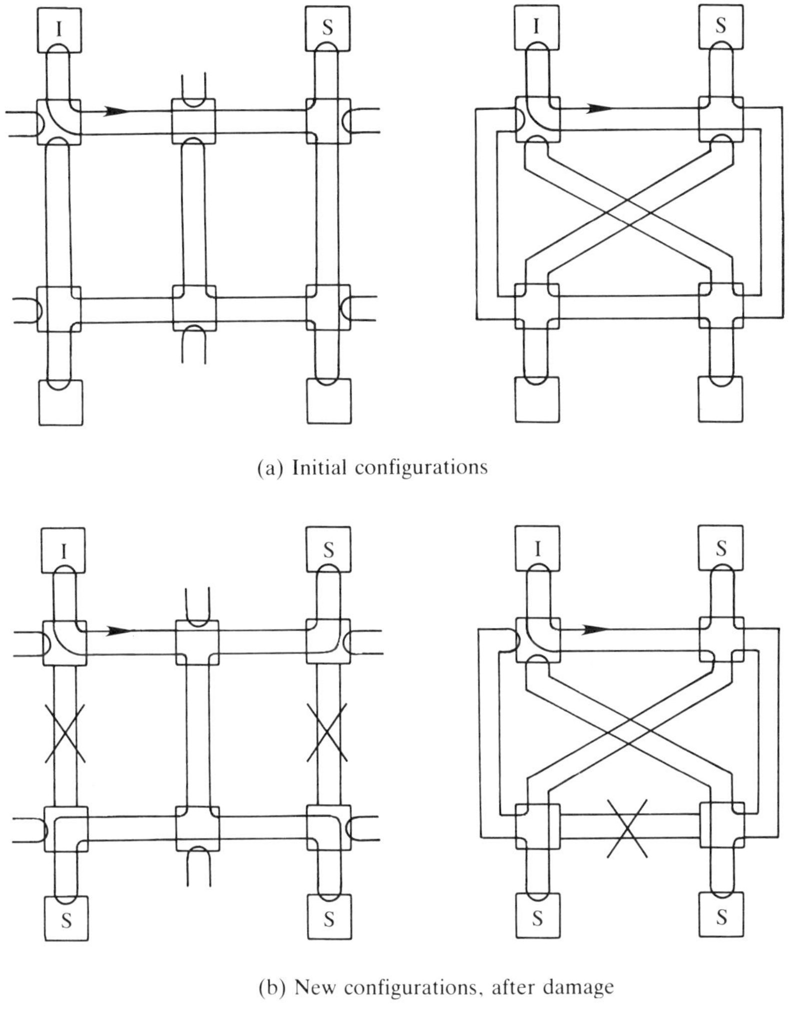

FIG. 9.2 Reconfiguration.

signalling to its repeater to start the reconfiguration process. Reconfiguration may also be started manually to enable controlled extension of the network to include new links and repeaters.

Figure 9.2(a) shows two ring configurations resulting from the reconfiguration procedure described above. In each case the initiating station is marked I and the other stations are marked S. In Fig. 9.2(b) the result of the reconfiguration of these networks after link failure is shown.

The reconfiguration scheme has the considerable advantage that it does not rely for fault tolerance on stand-by equipment taking over when a fault is detected. Stand-by equipment is apt to remain unused and thus untested for long periods, so that its failure is only detected at the time it is required to function. Since all working equipment is included in the ring in the GEC scheme, it is continually 'under test' and no special arrangements are required for testing back-up cabling and circuitry.

Prototype intelligent repeaters were built to work at a nominal rate of 2 Mbit/s. The repeaters each had three network ports and one station port. These ports used Manchester encoding, in which data and clocking information are contained in one signal, to drive the links. In order to function correctly, the total delay in any ring must correspond to an integral number of bits, so that the transmit and receive clocks at any station are in phase. The delay in a ring is fixed during normal operation but changes as links and repeaters are added or removed from the ring. Two techniques can be used to accommodate changes in ring delay, either the bit frequency can be altered or a variable delay can be inserted: the former was used in this network. For the reconfiguration procedure, the clocking and data information had to be separated at each repeater port. This was carried out using a type 1 phase-locked loop (Gardner, 1979) in each port, which introduces a frequency-dependent phase shift that allows changes in ring length to be accommodated with only small changes in bit frequency. The final stage of configuration involves the closure of the ring by the initiating repeater. Although the bit rate of the first outgoing signal and the final incoming signal at this repeater will be identical, they will in general not be in phase. To overcome this problem, a phase-locked loop with a long time constant is introduced into the clock signal at this ring closure point. This adjusts its operating frequency slowly to achieve zero phase difference between its input and output, the frequency adjustment being made sufficiently slowly to ensure tracking by the other phase-locked loops in the network. Each repeater must contain such a long time constant loop, but it is always arranged that only the loop in the initiating repeater is contained in the ring.

Each repeater has a probe signal generator which is used by the initiating repeater to produce an initial continuous Manchester signal

$$(010101\ldots),$$

which is used as the probing signal, and the startup signal

$$(1111000011110000\ldots)$$

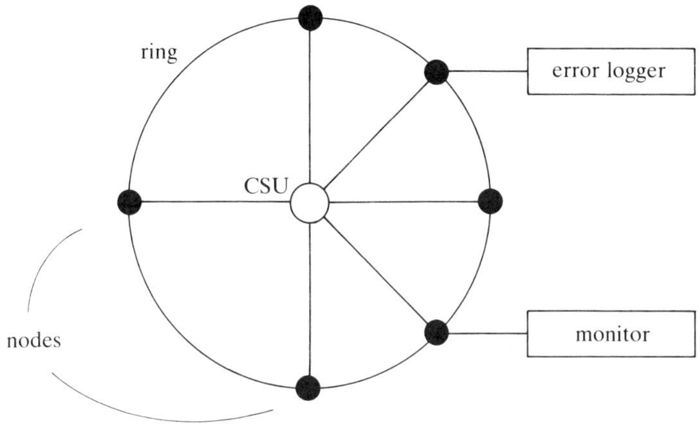

FIG. 9.3 The ring-star scheme.

which follows. Each repeater also contains a configuration controller, which implements the configuration algorithm.

The prototype repeaters built by GEC each occupied two and a half double Eurocards. Four repeaters were built and most of the experimentation during the lifetime of the OTS was concerned with testing the interworking of the token ring with the Cambridge rings. Subsequently, the repeaters were developed to a point where reliable reconfiguration could be demonstrated. These demonstrations were simply achieved by disconnecting one of the links in the configuration. Careful electronic design is required to achieve good ring stability and satisfactory transient behaviour after ring reconfiguration, since each repeater contains three phase-locked loops, rather than the single one of conventional repeaters. The design of suitably reliable electronics was not investigated since it was obvious that the circuitry would have to be incorporated into one or two VLSI chips for any commercial design. The repeater design was sufficiently well thought of for GEC to take out a British patent.

9.6.3 *The ring-star network*

Work at Loughborough concentrated on the development of a fault-tolerant scheme directly applicable to the Cambridge ring. The aim was to produce a design that would require only minor modifications to the range of ring equipment current at the time of the project. Also, since the maximum rate of transfer achievable between pairs of stations reduces as the length of the ring increases (see Chapter 2), a subsidiary aim was to keep additions to cabling lengths to a minimum. The design proposed by Loughborough (Chen, 1985) is built up from the standard ring topology with a star structure superimposed to provide the fault tolerance (see Fig. 9.3). In normal operation, data travels round the ring in the usual fashion. In the event of a failure in any repeater or link, the data path is diverted by relays to pass through the hub of the star, bypassing the faulty region of the ring. In the Cambridge ring, each node continuously checks the operation of the ring and sends a fault message to address zero if it detects any error. Using these messages an error logging station should be able to locate faulty components in the ring and signal this information to the central switching unit (CSU) at the hub of the star. A microprocessor in the CSU controls the operation of relays located both within the CSU and at each node and can switch these to alter the configuration of the network to isolate faulty ring components. The ring monitor station, which continuously checks the operation of the ring, will generally have detected the faulty condition and will certainly have noticed the loss of synchronization due to reconfiguration. It will then automatically attempt to re-synchronize the ring, obviating any requirement for special recovery measures. The scheme has the advantage that it can cope equally well with multiple failures.

The ring-star scheme in its simplest form is not really a practical proposition for a ring of any substantial size because of the cable lengths that would be required to connect the nodes to the star centre. The solution adopted was the obvious one of developing the scheme to allow a number of CSUs to be present in the network. These CSUs can be located near to the nodes that they serve. The arrangement pursued by Loughborough was an hierarchical one in which the CSUs themselves are in turn connected to a central CSU, called the master CSU (MCSU). The arrangement is shown in Fig. 9.4. A two-level hierarchy is shown but, for large rings, an additional layer of CSUs might be added.

A design for the CSU was produced by Loughborough based on the use of a Z80A microprocessor for control. The MCSU was a variation of this design with a different configuration of relays. The relay configuration for each node consisted of a relay on each side of the node and an associated relay in the CSU. Designs using only two relays are possible but in this case a link failure will cause the removal of a working node from the ring when the fault is bypassed. The original idea of using an error logging station to locate faulty components proved to be unreliable since, in some circumstances, nodes suffering a loss of power can generate incorrect error messages. The strategy adopted to circumvent this problem was to introduce an additional link between each node and its corresponding CSU. When a link is broken, the repeater immediately downstream generates an internal error signal. This signal is connected directly into an associated

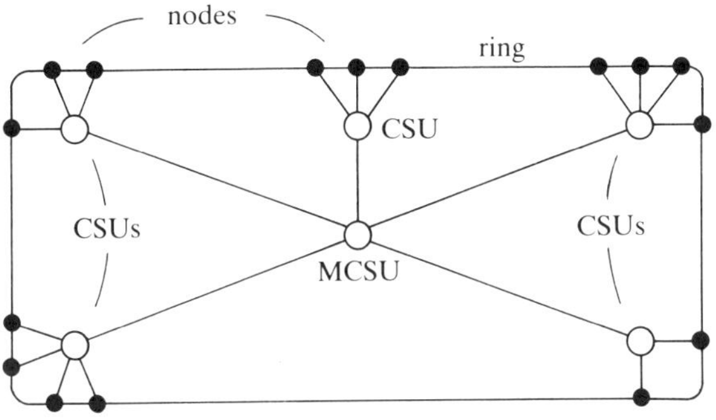

FIG. 9.4 Ring-star with two-level hierarchy.

latch in the CSU. The microprocessor in the CSU periodically reads these latches and can thus detect any ring break and locate its position.

Testing was carried out on a ring with six nodes using two CSUs and an MCSU. The system was able to recover successfully from both link and node failures, which could be induced artificially. Recovery from multiple component failure was also demonstrated. Using operator commands from a terminal connected to the error logging station, it was possible to control dynamically the configuration of the ring, removing parts of the ring and allowing the remainder to recover and continue operation. Such a facility would be valuable for installation and maintenance purposes.

The ring-star scheme has a number of attractive features: it adds fault-tolerance with minimal change to standard ring equipment and configuration; it can survive multiple faults; and it supports maintenance of the network. The scheme does, however, require the introduction into the network of a substantial number of additional components. Moreover, in normal operational circumstances, these components will be in a stand-by state. Very high quality detailed engineering design and construction would thus be required to realize the full potential of the scheme to improve substantially the reliability of a ring LAN. For particular applications, investment in this engineering effort might be justified.

9.7 Bulk data transfer

One of the fundamental uses of any computer network is the transfer of data. Particularly in the case of large amounts of data, this is also an activity for which the intrinsic satellite delay is of little consequence to an end-user. This delay does however pose technical problems in providing a reliable (file) transport mechanism which can deliver an appreciable fraction of the basic bandwidth to the user. In the case of an application which needs to send identical copies of a large file to a number of sites, there is also obviously scope for exploiting the intrinsic broadcast properties of the satellite segment.

Apart from any intrinsic interest in such investigations, the cost of satellite segments is such as to provide ample encouragement to optimize their use. While the experimental file transfer mechanisms reported here had particular relevance to the file distribution system described in Chapter 7, they were also directed towards obtaining a basic set of measurements, both to complement some of the others made in the project and to provide a foundation for ideas about bulk transfer mechanisms.

9.7.1 *Point-to-point transfer*

The main problem to be overcome in providing reliable, fast data transfer over a satellite channel is the end-to-end round-trip delay of nearly a second. To achieve throughput, as much data as possible needs to be kept 'in flight', consistent with the ability to recover from loss. In the UNIVERSE network, data corruption appears as loss of data and, more importantly in the current context, so does (satellite) bridge congestion. Conventional flow control allows both the receiver and the network switching elements to throttle a transmitter. No mechanisms for the bridges to throttle either each other or the transmitter were provided in the UNIVERSE network and so the general approach was to make the transmitter subservient to the receiver.

Briefly, the mechanism used for a transfer, in which the sender plays an entirely passive role, is as follows. An LVC is established by the receiver, which then sends a series of requests for portions of the file. The sender transmits whatever is demanded as fast as possible: it is the responsibility of the receiver to demand only what it anticipates being able to take. The sender maintains no state about the progress of the transfer. Each portion of the file requested is identified by its size and position in the file: it is called a 'gulp' here. Errors amount to all or part of a gulp being lost. The size of a gulp is a compromise between keeping as much data 'in flight' as possible and maintaining a high probability of the gulp arriving correctly. Loss of some of the constituent blocks of a gulp is generally caused by congestion in a satellite bridge and necessitates retransmission of the lost parts.

Measurements were obtained for two forms of procedure, referred to as single and double 'gulping'. These procedures are analagous to conventional single and double buffering schemes, the buffer in this case referring to a gulp. Clearly, the single gulping scheme, in which a receiver requests a gulp and awaits its arrival before requesting another, requires large gulps (and hence substantial buffering in the recipient) to achieve good performance. This is also very likely to cause bridge congestion. Substantial improvement, in terms of increased throughput for substantially smaller gulp size, can be obtained with a double gulping strategy, in which the transmission of one gulp is overlapped with sending the request for the next gulp.

Measurements were made of memory-to-memory transfers between large servers at two different sites. Most of the measurements were repeated between systems on the same ring, for comparison, to eliminate effects not due to the satellite segment. The time taken to transfer various amounts of data using different sizes of gulp was measured. Basic blocks of maximum size (2048 bytes) were used throughout. For the purpose of obtaining basic measurements, cases where loss and retransmission of data occurred were excluded from any consideration other than indicating the onset of bridge congestion.

9.7.2 *Simple gulping model*

In order to relate the timing measurements made in this experiment to others made elsewhere in the project, a simple model of the gulping process was used. The relationship of gulping parameters to both the satellite bridge behaviour and the performance of the ring interface used in the large server were of particular interest.

All the measurements were of the time T taken to make a transfer consisting of N gulps, each of size S blocks. N and S were varied and the transfer time was measured from the start of transmission of the first request to the end of the arrival of the last data block. Generally, T is assumed to consist of a fixed transfer overhead time, T_a, and a time proportional to the number of gulps. The time to obtain a gulp was also considered to consist of two parts: one part independent of the size of the gulp, called the gulp overhead, T_{go}, and the other, T_g, proportional to the size of the gulp. Thus for single gulping, the basic relation is

$$T = T_a + N(T_{go} + T_g). \tag{9.5}$$

T_{go} is the time from the request for a gulp being sent to the start of reception of the gulp; T_g is the time taken to receive the gulp, from the start of the first block to the end of the last. T_a, the transfer overhead, includes the transmission of a closing acknowledgement but not connection establishment time; T_a was negligible for ring transfers. For satellite transfers it was essentially dominated by the time to acquire a large window allocation in the satellite TDMA scheme (see Chapter 4), which would be obtained as a result of sending the first data block. This 'acquisition time' is expected to be about 1.2 s (Chapter 4; Waters, 1984; Report 17). This is quite consistent with the measurements. Since the errors in its determination in this way are large, it was treated as known input to this experiment. T_{go} is predicted to be about 810 ms from the satellite round-trip delay measurements given in Waters (1984); as with T_a, this is reasonably consistent with the results but the errors are quite large. The time taken to receive a block, T_b, was also used, defined essentially by the relation

$$T_g = ST_b. \tag{9.6}$$

Combining (9.5) and (9.6) gives

$$T = T_a + NT_{go} + NST_b. \tag{9.7}$$

In double gulping, requests for data, travelling from the receiver to the sender, are overlapped with data travelling in the opposite direction. The over-simple procedure used in the experiment was to begin by sending two requests and thereafter to request further data immediately after receipt of each gulp. This initiated a pattern in which pairs of gulps arrived in close succession, with the first of a pair being requested between the two gulps of the previous pair. If T_{go} is assumed to be larger than T_g, that is, it takes longer to request a gulp than to receive it once it has begun to arrive, then it follows that the reception time of the second gulp of a pair is entirely overlapped with the time for the next request, except for the final gulp. For an even number of gulps

$$\left.\begin{aligned} T &= T_a + T_g + N(T_{go} + T_g)/2 \\ T &= T_a + NT_{go}/2 + (1 + N/2)ST_b \end{aligned}\right\} \quad (T_{go} > T_g), \tag{9.8}$$

where the second form is obtained from the first by using (9.6).

There is an alternative regime which occurs when the receiving LDC is CPU limited (Chapter 4; Waters, 1984) and the data blocks are held up in the LDC for sufficiently long that the average time to receive a gulp exceeds the gulp overhead time, T_{go}. Under these conditions, the following two simpler expressions apply:

$$\left.\begin{aligned} T &= T_a + T_{go} + NT_g \\ T &= T_a + T_{go} + NST_b \end{aligned}\right\} \quad (T_{go} < T_g). \tag{9.9}$$

9.7.3 *Results*

Full details of the measurements and analysis made in this experiment may be found in Report 16; only the more salient points are noted here. Figure 9.5 summarizes the behaviour in graphical form. Transfer times for

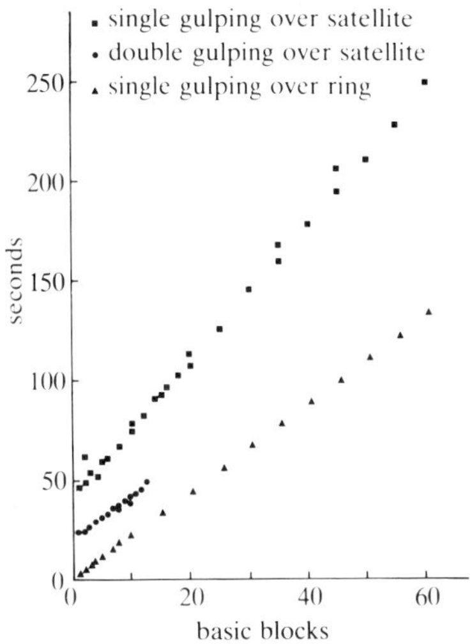

(a) *Transfer time (T) vs. gulp size (S); transfer size (N), 50 gulps.*

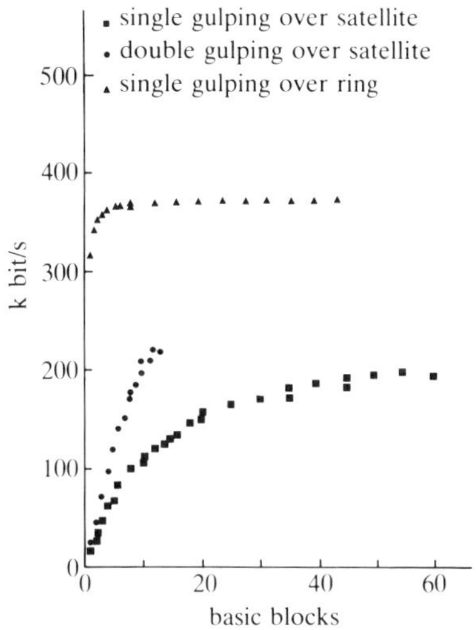

(b) *Rate vs. gulp size (S); transfer size (N), 50 gulps.*

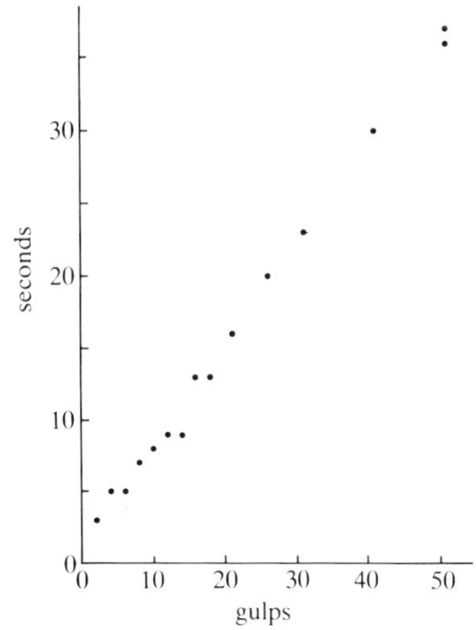

(c) *Transfer time (T) vs. transfer size (N); gulp size (S), 8 blocks.*

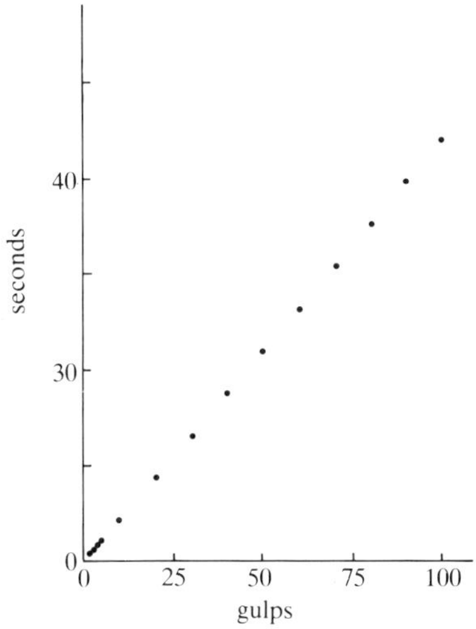

(d) *Transfer time (T) vs. transfer size (N); gulp size (S), 10 blocks.*

FIG. 9.5 Gulping performance.

various sizes of gulp are shown in Fig. 9.5(a); the corresponding rates are given in Fig. 9.5(b). Transfer time as a function of transfer size for double gulping over the satellite is shown in Fig. 9.5(c); Fig. 9.5(d) shows the corresponding behaviour for transfers between two systems on the same ring. Measurements given in Report 16 but not reproduced here showed, as expected, that there is almost no difference between single and double gulping for ring transfers. The slopes of the ring transfer timing measurements shown in Fig. 9.5(a) and Fig. 9.5(d) yield a value for T_b of about 42 ms, giving an effective transfer rate of some 375 kbit/s. (The expressions for ring transfer times may be obtained from those for satellite transfers by setting T_a and T_{go} to zero in (9.5) and (9.7) or (9.9).) This is in agreement with other measurements of the basic block performance of large servers in the two-slot ring in use at RAL.

Figures 9.5(a) and 9.5(b) demonstrate the substantial improvement gained by using double gulping. The maximum rate reached with single gulping was about 200 kbit/s, for which gulps of 120 Kbyte were needed. Using double gulping, a rate of about 220 kbit/s was achieved but gulps of only 24 Kbyte were required.

Careful analysis of the results indicates that the satellite bridge is CPU limited during double gulping for $S = 12$ and is exhibiting signs of limiting behaviour for $S = 10$. Certainly (9.9) applies for $S = 12$. For gulp sizes greater than this, blocks are regularly discarded by the bridge because of congestion.

If the various overheads (T_a, T_{go}) are eliminated, then both single and double gulping measurements yield effective figures of 240 to 250 kbit/s for the basic block transfer rate within a gulp. This is to be compared with the limiting rate of 270 kbit/s for the LDC receiving from the satellite (Chapter 4; Waters, 1984). The cause of the 10–12 per cent discrepancy is not known, though one may speculate that it is in part due to the bridge handling gulp requests, which was not required in the LDC measurements.

9.7.4 *Comments*

Largely through lack of effort, relatively little work was done towards constructing a reliable protocol for bulk transfer based on these results. Versions for disk-to-disk transfer were made with simple recovery based upon asking for retransmissions of those blocks missing from a gulp, using the direct access nature of the media at each end to avoid the need to maintain sequentiality during recovery. However, no work was done on controlling the rate of transmission and, more particularly, on the need to reduce the rate in the presence of congestion. It is also clear that a more regular flow of data would be preferable to the bursty flow induced by the simple procedure used here.

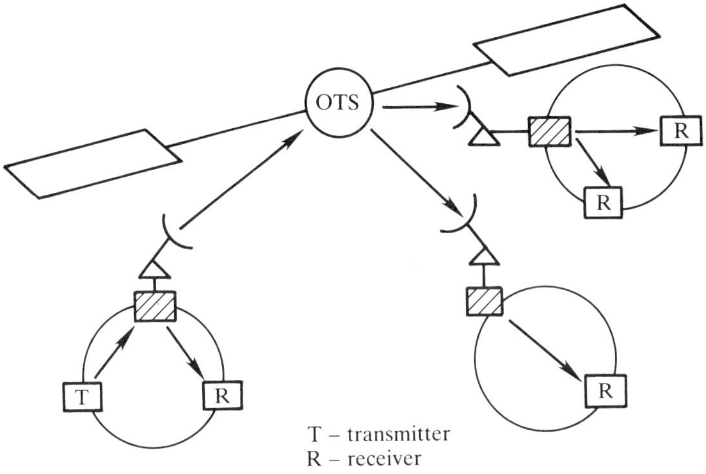

FIG. 9.6 Basic broadcast configuration.

More recently, Clark *et al.* (1987) have reported considerable progress on a high-throughput transport protocol called NETBLT, in which the emphasis has been on decoupling flow control and error recovery. The protocol is driven by the receiver which maintains the protocol state. Blocks (packets) within a gulp (buffer) are transmitted at a rate controlled by the receiver; error recovery is by the means suggested above but the relevant blocks are transmitted, like any others, at the rate currently in force. This work has demonstrated the success of this approach over the DARPA wideband satellite network by achieving almost 95 per cent efficiency at the maximum rate of 1 Mbit/s available to the NETBLT layer.

9.7.5 *Multicast transfer*

The primary goal of this experiment was to develop a mechanism for multicast file transfer which made efficient use of the satellite channel bandwidth. Speed of transfer was of secondary importance. As described in Chapter 3, the broadcast nature of the satellite channel was used to provide an inter-site broadcast service; this was combined with a set of 'broadcast servers' at each site to provide a multicast service which covered the whole network. Multicast file transfer was envisaged as a specialized activity for use only among agents such as those developed in the file distribution scheme (Chapter 7); clients were not in general expected to implement this protocol. The experiment was concerned primarily with the satellite segment and the basic configuration indicated in Fig. 9.6 was used.

The protocol is described in terms of the actions performed by the participants, of which one is the *originator* of a file and the rest are *recipients.* To set up the broadcast, the originator first opens a connection (LVC) to the broadcast service to transmit on the static file broadcast channel. If this is successful, it then opens a *control* LVC to each recipient, specifying the destination file name, the file size, and the size of blocks to be used in the transfer. A recipient, correspondingly, opens an LVC connection to its local broadcast receive service and then accepts the control connection from the originator. Each recipient's control connection to the originator is used to request data and to terminate participation; the volume of data flowing on such a connection is quite small. As an aside, it may be remarked that in the case of receive-only earth stations, it would be quite reasonable to route these connections over quite slow terrestrial links, while broadcasting the bulk data over the satellite channel.

The transfer is driven by requests from the recipients. The originator receives requests but delays servicing them in an attempt to synchronize the recipients. The originator's model of the transfer is contained in three quantities: the number of recipients, n; a file position, p, which is initially zero; and a value k described below. If the originator receives a request for data preceding p, it broadcasts this data immediately. When it receives a request for data beyond p, the originator may delay broadcast. It notes the number of recipients, k, who have made such requests and how far beyond p each has requested.

The originator uses a delay function, T, dependent on the difference $n - k$, such that T is monotonically increasing and $T(0)$ is zero. The timer starts as soon as the first request is received but is modified by subsequent events. The delay function used in the experiments was

$$T(n-k) = 2(n-k) \qquad \text{(seconds)}.$$

Thus, if all recipients have requested beyond p, the originator broadcasts immediately. When it decides to broadcast, the originator transmits data to fulfil the request of the recipient asking for the least data. Each block sent by the originator contains its file position and length. When a broadcast is made, p is increased by the number of bytes in the broadcast. The originator will cease operation when it deems all recipients to have finished, either by receiving explicit notification on the control channel or through time-out.

Each recipient performs a complementary algorithm. It has a number of buffers, which it cycles along the file. Since both source and target files are required to be randomly accessible there is no requirement for requests to be sequential. Consequently, there is no window as such. For each buffer, a request is made for a block of data (including position and size),

TABLE 9.1 Broadcast protocol throughput (kbit/s). (© *1984 ACM*)

Buffer size (Kbytes)	*Number of recipients* *1*	*2*	*3*
4	27	26	24
10	60	52	50
20	71	56	52

starting at the beginning of the file. Requests for contiguous data may be merged by the originator and treated as a single request. A typical request would be for 20 Kbyte. When requests are fulfilled, data is written to the file and a new request is made for the next unrequested portion of the file. Whenever a request times out, it is repeated. This process continues until either the entire file has been transferred, in which case the recipient notifies the originator, or the recipient times out the originator.

This procedure proved to be quite efficient in its use of the satellite channel; retransmissions of any sort were rare, in particular those due to the presence of multiple recipients. Table 9.1 shows the results of some measurements made of broadcasting to one, two, and three recipients. Performance was limited by disk access for the larger block sizes (see previous sections). There is degradation as the number of recipients increases, partly because the chance of interfering with other data is higher. Not suprisingly, results showed that the variance in throughput increased as the number of recipients increased. This was best illustrated by the few results taken with four recipients, using a 10 Kbyte buffer: some results were as high as 48 kbit/s, while others were half this speed. Results for four recipients are not shown in Table 9.1, as they were incomplete, owing to a period of low reliability in several of the earth stations. Nevertheless, these initial tests were quite encouraging; it would be interesting to investigate using a larger number of receivers and to study the relationship of this mechanism to NETBLT principles.

10. Demonstrations

10.1 Introduction

Early in the life of the project, it was suggested that it would be appropriate, in view of the nature and funding of the project, to mount public demonstrations of the techniques and ideas being developed. This was at a time when the government was mounting several initiatives aimed at increasing public awareness of the nature and application of information technology. In particular, the Department of Education and Science (DES) had begun its programme of funding microcomputers for schools. The British Broadcasting Corporation had chosen a microcomputer (to be known as the BBC microcomputer) developed by Acorn Computers Ltd to be used in television programmes about computer literacy and, amongst others, this would also be used in the DES programme. Furthermore, the DTI was planning a series of activities for 1982 as part of its information technology awareness initiative, called the IT82 programme. Public demonstration had not been envisaged when the project was proposed and there was consequently no allocation of effort or resources for this purpose. It was recognized that it was appropriate for a project of this scale, funded with public money, to mount public demonstrations of the technology it was developing, but that to do so required additional funds and effort.

In order to make this possible, the DTI provided funds in the region of £150k, which covered all the direct costs associated with mounting demonstrations at public exhibitions. These included hiring additional effort; exhibition attendance costs, including stand construction and publicity; and various incidentals, such as transportation. The additional effort was obtained through a contract for two further people from Logica, who provided the overall management, organization, assembly, testing, and transport for the exhibitions. The technical development, mounting, and demonstrating effort needed for each individual subsystem was donated by the participants. The endeavour was managed by two committees: a small management group and a larger technical working party, responsible for agreeing and progressing technical developments.

10.2 The exhibits

The project mounted demonstrations at the exhibitions listed in Table 10.1. The first two, held at the Barbican in London, were concerned with appli-

TABLE 10.1 Exhibitions visited by UNIVERSE

Title	*Place*	*Dates*
Info '83	Barbican, London	22–25 Feb. 1983
Business Telecom	Barbican, London	24–26 May 1983
World Telecom '83	Palexpo, Geneva	26 Oct. – 1 Nov. 1983

cations of telecommunications and information technology in the context of general business use. These exhibitions were similar, with most of the stands showing current or imminent products. World Telecom is the major exhibition for the telecommunications industries of the world and is held in Geneva every four years. It is a very large exhibition, with displays of products and research activities covering all aspects of communications. For World Telecom '83, the project demonstration was part of the UK exhibit organized by BT.

The theme of the project demonstration was to show possible ways in which new, high-speed forms of communication could in the future support integrated applications using a variety of media. The exhibit took the form of an additional temporary site added to the UNIVERSE network. It was fully equipped with a satellite earth station, satellite bridge, and Cambridge ring which ran through the exhibit stand. Attached to the ring was the usual infrastructure equipment (nameserver and bootserver) together with terminal equipment on the stand used for a number of live demonstrations of communication with other sites.

As a means of making what was a technical and highly experimental demonstration more accessible to the viewer, considerable use was made at all three demonstrations of the BBC microcomputer. At the time of the project, this computer had only recently been launched, amid considerable publicity, and had sold in substantial numbers. From the standpoint of the project, it provided a cheap, programmable keyboard and colour display, whose bus was directly interfaced to the Cambridge ring.

Broadly speaking, the format of the demonstrations at each of the exhibitions was similar. Not all of the individual demonstrations were shown at every exhibition. Moreover, development was continuous and later exhibitions benefited from better versions of some of the systems. The approach taken in the remainder of this chapter is to describe the overall format and then to describe the latest version of each demonstration, without reference to the particular occasions on which it was exhibited.

The satellite link with the other sites was provided by a mobile earth station on loan from BT. As could be expected, at none of the exhibitions was it possible to locate the earth station close to the main exhibition stand. At the Barbican, the nearest site for the trailer was the grounds

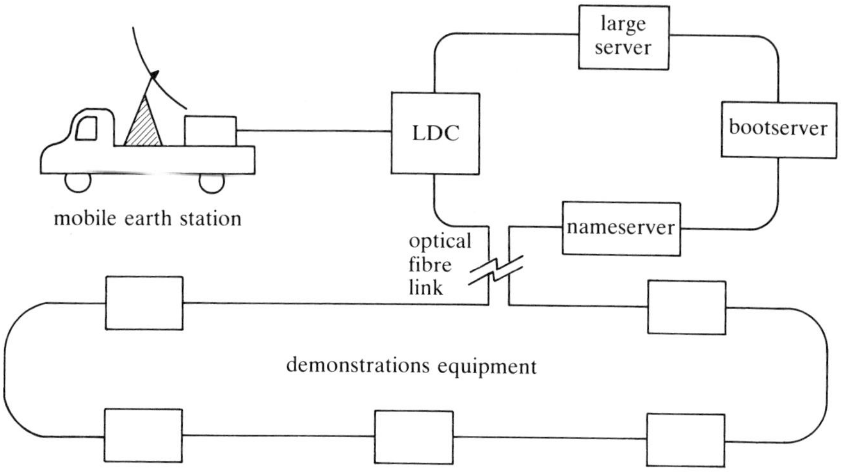

FIG. 10.1 General configuration of equipment for exhibitions.

of the Honourable Artillery Company, who were kind enough to extend permission for this unusual use. At Geneva, there were both internal and external demonstration sites. Space was at a premium and some of the demonstrations were mounted in a mobile exhibition container placed outside, beside the satellite earth station. In all cases, the equipment was arranged in two clusters, separated by up to 1 km. The ring used to link all the demonstrations to the satellite bridge was extended by use of two optical fibre segments (Chapter 5) to link the two clusters. The LDC of the satellite bridge was located close to the earth station, avoiding the need for extended radio or intermediate frequency links. The optical fibre links worked well in all three instances and, in particular, demonstrated their immunity to interference. Having linked the two clusters in this way, the precise placement of the various demonstrations had logistical implications but none from a communications standpoint. The general arrangement is shown in Fig. 10.1.

As a point of general strategy, most of the demonstrations sought to arrange that backup or alternative configurations were available, so that some form of demonstration could be achieved during periods when the satellite was scheduled not to be available or in case of equipment failure. Two approaches were used, often in combination. For those demonstrations making use of a local client system communicating with a remote server, a local server was also provided, where possible. Furthermore, a number of demonstrations used common equipment, so that spares could be reduced and, in case of further failures, demonstrations could be scheduled to allow equipment to be used in more than one demonstration.

10.2.1 *Slow-scan video*

As a part of the experimental programme, Loughborough developed a number of program-controlled framestores which were interfaced to the network. This work is described in Section 8.3. In their final form, these framestores could typically handle colour pictures of 512×256 pixels with 16 bits per pixel. Transmission rate was limited to 220 kbit/s or about 9 s for a single frame.

The basis of the demonstration was to site four framestores on the network, two at the exhibition and one each at Loughborough and RAL, as shown in Fig. 10.2. Attached to each framestore was a camera and display monitor. A BBC microcomputer was available on the stand and could be used to control which of several frames in a framestore was displayed on the monitor; it could also instruct a framestore to renew one of its images, either by 'grabbing' whatever was the current image captured by the framestore camera or by transferring an image from one of the other framestores. An additional facility was also available, whereby one framestore would choose to display a particular image which it would continuously update from another framestore, which was itself continually grabbing a new frame between transmission of successive images.

In order to add interest and attraction to the demonstration, some simple servo-mechanisms were designed, built, and interfaced by Loughborough to a BBC microcomputer. These enabled the pan, tilt, and zoom of a remote camera to be controlled from another BBC microcomputer situated anywhere on the network. Following the same general principles, a stationary robot arm was also interfaced via a BBC microcomputer to the network. From a demonstration presentation point of view, these additional components were a success. At Telecom '83, it was possible from Geneva to use the robot arm in Loughborough to move pieces around a draughts board and to observe something of what was happening via the slow-scan video link. The general arrangement of this control equipment is included in Fig. 10.2.

The use of such an arrangement to control a remote manipulator was not regarded as significant: it simply added an element of amusement to the demonstration. However, it was observed that the ability to control the remote camera, even by so cumbersome a method as typing commands at a keyboard, increased considerably the usability and range of potential application of such a system. This is borne out by practical security surveillance systems, for example; other possible applications might include remote discussion and assistance for problem determination and repair, and also remote learning.

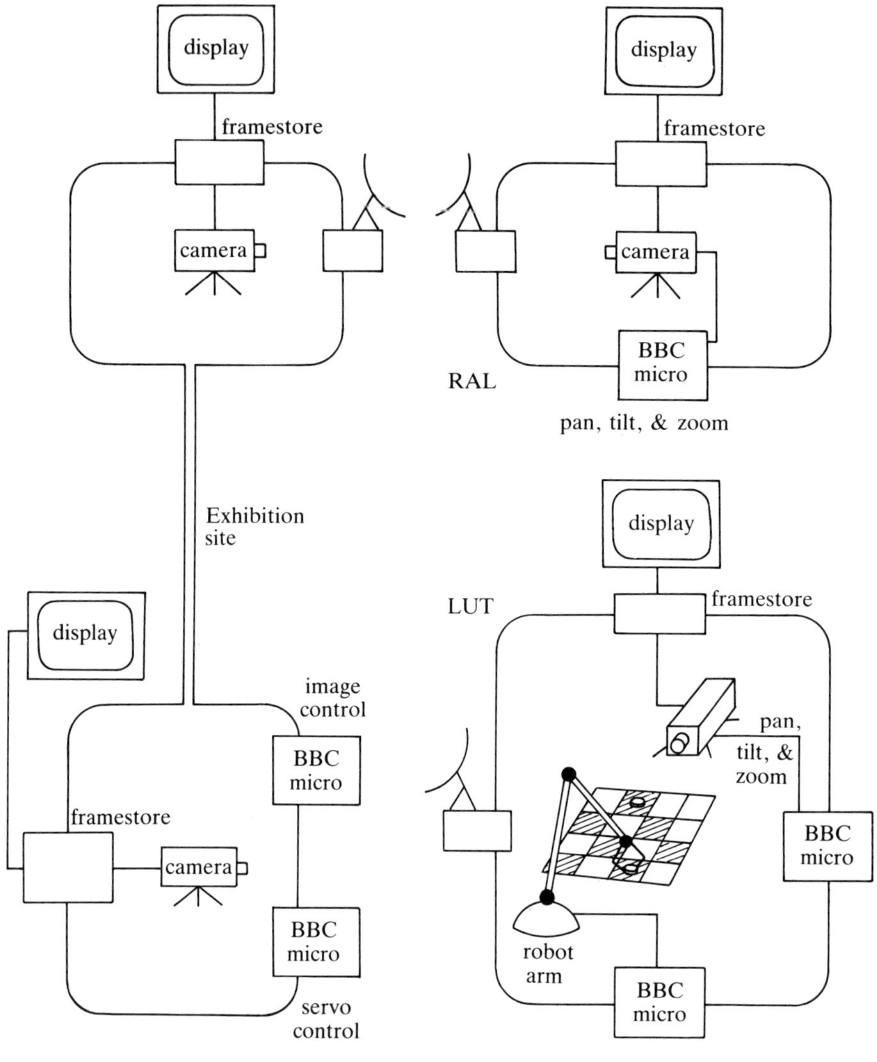

FIG. 10.2 Slow-scan video and servos.

As was mentioned earlier, the satellite link was not available continuously throughout the period of the exhibition and it was generally arranged that demonstration was possible locally. For this purpose, two framestores with associated cameras and display monitors were made available at the exhibition.

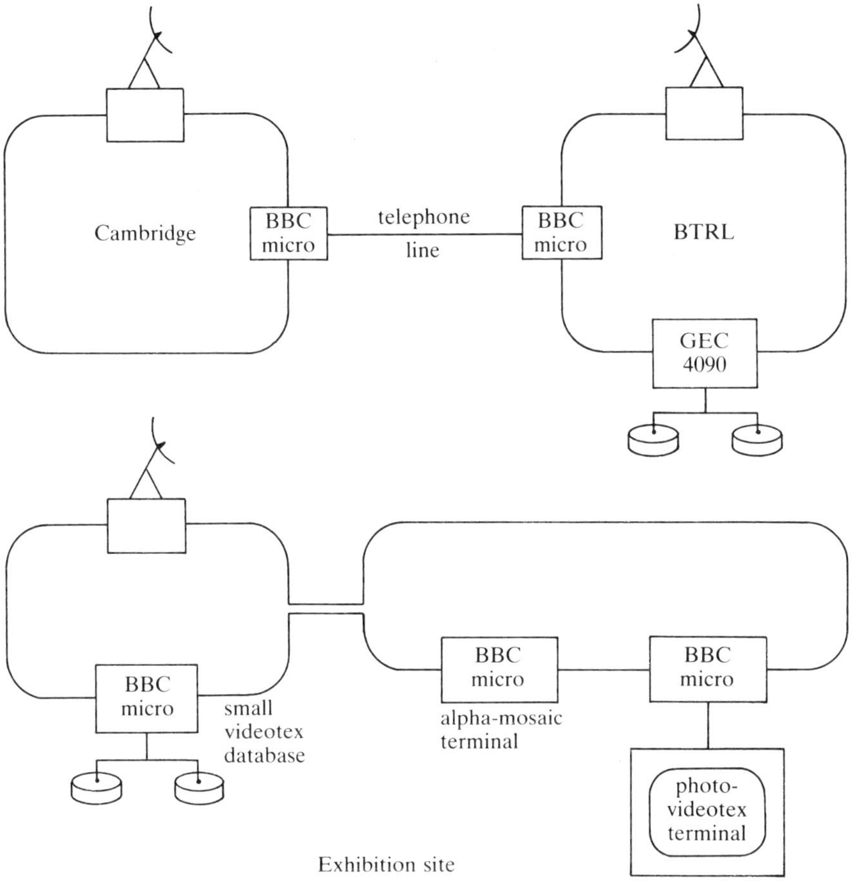

FIG. 10.3 Interactive videotex.

10.2.2 *Videotex*

The experimental programme included some investigation of the operation of *interactive videotex* (hereafter called simply *videotex*) over the UNIVERSE network; this is described in Section 9.2. As part of this work, BTRL developed several videotex clients and servers which were attached to the network and were demonstrated at two of the exhibitions (Business Telecom and Telecom '83). The general arrangement of equipment for the demonstrations is shown in Fig. 10.3. Access via an alpha-mosaic terminal was shown at both exhibitions, whereas photo-videotex access was only demonstrated at Telecom '83. It had originally been hoped to demonstrate the advanced (full-screen) photo-videotex terminal at Geneva but the software could not be completed in time. Instead, a photo-videotex terminal with less advanced capabilities (alpha-mosaic with photo-insets up to one sixth of total screen area) was shown.

Two videotex database servers could be accessed. The main one was situated at BTRL and consisted of a GEC 4000 executing Prestel software modified to work under OS4000 and to handle terminal access via the ring (Chapter 9). The second server was a small one, based on the BBC microcomputer. The implementation was a simple one to allow basic tests to be done quickly and easily. It was also portable, so that the system could be taken to other sites for test and demonstration purposes: in the case of the latter, it was used as a backup server and situated at the demonstration site. The server was interfaced to the ring in the usual way for BBC microcomputers (Chapter 6) and used two double-sided floppy disks to provide 800 Kbyte for database storage. A basic alpha-mosaic videotex frame can be stored in 1 Kbyte; photo-videotex frames require additional storage of up to 8 Kbyte. For demonstration, 5 photo-videotex and 25 alpha-mosaic frames could be accommodated on a single disk. The performance achieved after tuning the BBC microcomputer server was more than adequate: indeed, some reduction in the data transmission rate was necessary to accommodate the advanced photo-videotex terminal.

A number of problems beset this demonstration. Throughout the project, difficulties were experienced by BTRL in obtaining the requisite performance from their satellite earth station, which was of different design from the others. To safeguard the demonstrations, BTRL developed, with the aid of Cambridge, a primitive point-to-point link which used two BBC microcomputers and a 4.8 kbit/s (later 9.6 kbit/s) public switched telephone connection between BTRL and Cambridge. For Business Telecom, this was the sole means of access since the BTRL earth station had not been commissioned. At Geneva, though the earth station was operational, the difficulties of its use were exacerbated when, owing to problems of FDMA loading on the satellite channel, the project was instructed to reduce transmit power (from all sites) to the satellite, thus substantially reducing the signal to noise ratio: this caused difficulties on all the UNIVERSE links and severed direct communication with BTRL altogether. Thus the link between Cambridge and BTRL, intended only as a fall-back measure, was in fact used for the majority of both demonstrations. In spite of these various difficulties, BT were able to operate videotex over the network sufficiently well to produce an interesting and attractive demonstration.

10.2.3 *Multi-media document retrieval*

The existence of digital encodings not only for conventional text and graphical images but also for speech and raster images, coupled with signs of the

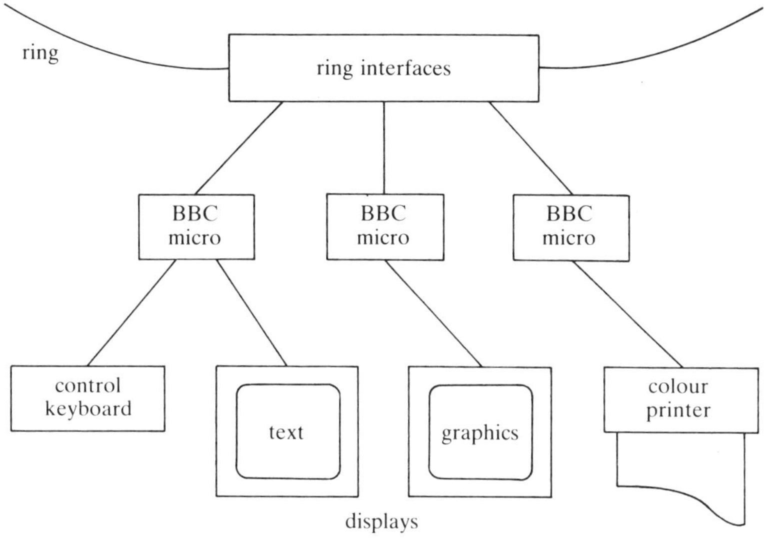

FIG. 10.4 Model demonstration multi-media terminal.

approaching emergence of both sufficiently dense direct access storage media and appropriate forms of high-speed digital communication, all suggest that the 'multi-media document' may become a practical possibility. What such an object is, how it is structured, transferred, and accessed are all topics of investigation. It is assumed here, at least for the purpose of access, that such a document may be regarded as being made up of a number of segments and that each segment is homogeneous with respect to medium and encoding.

To access such a document, whether for composition, editing, or simple retrieval and replaying, a multi-media workstation is obviously needed. One approach to providing such a station on an experimental basis is to use a LAN to provide an extremely low, but correspondingly flexible, level of integration among a set of devices, each capable of handling a single medium, and a conventional workstation acting as the co-ordination and control interface.

The model multi-media terminal developed by UCL for demonstration was arranged on these principles (see Fig. 10.4). The terminal was constructed out of a set of three BBC microcomputers, each interfaced directly to the demonstration ring. All communication, both among the component microcomputers of the terminal and with the database servers, was by SSP transactions. The principal microcomputer acted as the control interface and also handled the display of text segments. Specified parts of documents could be retrieved from a choice of several database services, situated either on the local ring or at UCL. The two remaining microcomputers acted essentially as slaves under the control of the principal: one was used to display colour alpha-mosaic pictures or diagrams (in videotex format) and the other was used to provide colour printouts of specified parts of the retrieved documents, which could be handed to visitors.

10.2.4 *Stored voice*

As will be evident from the description in Chapter 8 of the work undertaken on packet voice in the project, the main thrust in this area was the investigation of a packet voice protocol capable of supporting a conventional telephone conversation. A subsidiary, substantially simpler undertaking was the development of a demonstration stored voice message system, which could be viewed as an extension of electronic mail to include spoken messages.

A central message store was implemented on a GEC 4000 system, as a subsystem executing under the control of the standard operating system, OS4000. A single file of some 8 Mbyte, sufficient for about 1000 s of speech, was set up as the message store. This could be used for up to 200 individual messages. The coded dialling system used to access the service is described in Chapter 8 (Table 8.1). Upon completion of recording a message for a nominated recipient, the system sent an electronic mail message (by use of a table relating telephone numbers and mailbox addresses) to inform the recipient of the new message and its identity. The recipient could choose either to listen to all new messages destined for the designated telephone number or to a single message specified by its identity number. Messages were deleted on the basis of age when the store became full and the space was needed.

Although towards the end of the project, the packet voice protocol properties were sufficiently well understood and the Z80 telephone systems (Chapter 8) were provided with code which could sustain either conversational use or voice store access, in practice at all the demonstrations simpler systems were used. To overcome the effect of statistical fluctuations in the transit time, 2 s (16 Kbytes) of buffering was introduced into the Z80 receive logic. This figure was found experimentally to be large enough to provide perfect speech reproduction. Details of the components of the Z80 telephone controller and its connection to the network are given in Chapter 8; a typical network configuration which includes access to a voice message store is shown in Fig. 8.2 . There was no provision to enable the operator to control which voice store was accessed: this was determined by whatever was held in the local nameserver under the entry for the name *voice*.

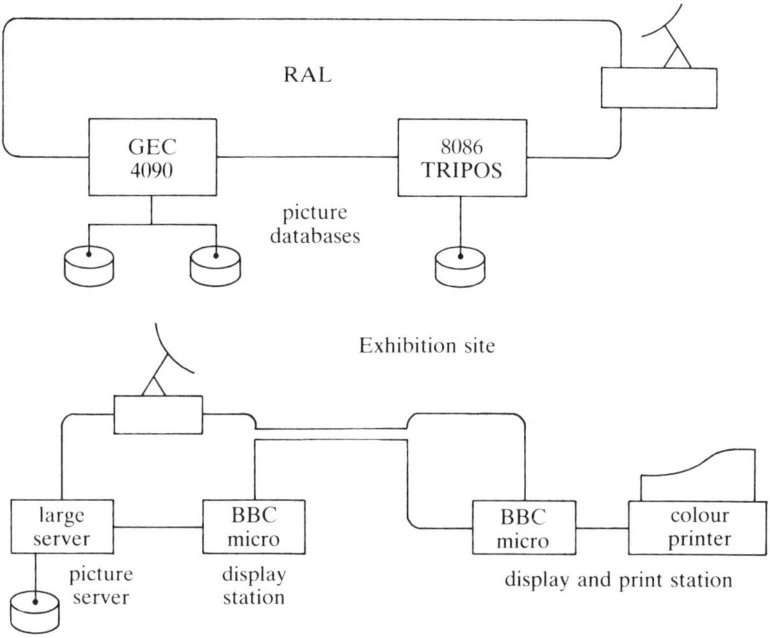

FIG. 10.5 Image retrieval at Telecom '83.

The main recorded-voice server was implemented on a GEC 4090 at RAL during 1982 and remained in operation as an experimental service for the remainder of the project. Two other such servers were in operation at various times during the project, both on GEC 4065 systems. One of these was on loan from GEC Computers Ltd for use at the first two exhibitions. The other was at the Marconi Research Centre. At those exhibitions where a 4065 was available, the demonstration continued to be available even when the satellite was not.

From a technical standpoint, this demonstration was satisfactory in that it showed that given a network of this type, all the technology is available to make simple applications based on communication with a central message service. At the technical level there appears to be no obstacle to the integration of this sort of system as a component in a distributed office system. Indeed, the form taken by the experimental system demonstrated a minimal form of integration with electronic mail.

10.2.5 *Stored image retrieval*

The final demonstration showed a very basic form of picture information retrieval from one or more remote databases. The familiar BBC microcomputer functioned as the access terminal and the user could choose which service to access and which frame (picture) to retrieve. The images were a mixture of text, graphics, and bit-map images, the majority being the latter. A square subframe amounting to 80 per cent of the BBC microcomputer's overall display area was used: a bit-map image of this size required 16 Kbyte of data or eight full-sized basic blocks to replenish. Although this system only provided low resolution and a limited colour repertoire, quite interesting pictures with recognizable detail could be displayed.

The general arrangement of components for the demonstration, as they appeared at Telecom '83 at Geneva, is shown in Fig. 10.5. The databases used for all the exhibitions were situated at RAL and at the exhibition site itself, to allow the demonstration to continue in the absence of the satellite segment. Several different servers, based on both TRIPOS and GEC OS4000 operating systems, were capable of providing the database services. Since the demonstration was one which gave some impression of the speed with which the network could convey information, even to a very cheap display station, some effort was expended to ensure that a picture database server was always available. This was particularly the case for the first exhibition, since a government minister had been invited to 'open' the network formally and it was by retrieving an image from one of the servers that he achieved this. For later exhibitions, UCL enhanced the basic system developed by RAL by adding the ability to produce colour prints of the retrieved image, which could be given to interested visitors.

Without modification, this demonstration may be considered to be very similar to a miniature photo-videotex application, with low-resolution images. Integrated into some larger system, it could constitute a model component of a multi-media document retrieval sysem. For exhibition purposes, it served to demonstrate that a simple personal workstation could effectively use the network to access quickly pictorial information from databases anywhere on the network.

10.3 Conclusion

The whole operation of mounting demonstrations was one which was not viewed with unanimous favour by all participants. The UNIVERSE project was originally proposed and funded for the purpose of research and development. The suggestion that public demonstrations should be given came from elsewhere and it was recognized at the outset that even with additional resources for exhibition purposes, it was inevitable that effort funded for research and development would be diverted to develop and mount the demonstrations. Agreement to the suggestion arose from a number of factors. The funding was substantially from public sources, primarily the DTI; the project as a whole was regarded as being loosely related to the IT82 programme; and the DTI was prepared to meet the additional costs in order that the project could contribute to IT82 by demonstration. Furthermore, the private funds were provided by the industrial collaborators and the possibility of public demonstration offered the opportunity for public exposure of both new products and pre-competitive prototype development.

In retrospect, it is possible to pick out a number of effects which the public exhibitions had upon the project. A major, positive effect was to impose a very definite deadline (February 1983) by which the network and a substantial number of subsystems had to work. This had a marked effect on equipment and systems being made to work: the toll in terms of frantic effort and long hours was also evident! As was foreseen, quite a substantial diversion of effort at several academic sites did take place. While it is hard to quantify the effect of this on the research programmes of the project and of individuals, it was clear that the effect was adverse and was unfortunate in this respect for the project.

In terms of meeting the overall goals of demonstrating the technology being developed by the project and contributing to the general awareness of the future possibilities for information technology, the demonstration programme appeared to be quite successful. On the stands at the various exhibitions, a fair degree of interest was shown by visitors and, at the technical level, most of the demonstrations managed to function for most of the time. That a whole site, with all its many components, could be moved and assembled and made to work so quickly while most of the equipment was of an experimental variety spoke well of the general principles underlying operation of the network. Finally, the project attracted a good share of press attention at the exhibitions and all of it was favourable. The participants themselves came away with the general view that while the effort had been considerable, the outcome had been satisfactory.

11. Conclusion

In the preceding chapters, all aspects of the project have been described and some assessments made of the achievements in particular areas. Here, it is our purpose to present a somewhat broader review of the whole project, consider its strengths and weaknesses, and put it in context with other developments which have taken place elsewhere. We also indicate a personal view of more recent pertinent developments, suggesting some possible directions for future investigations.

11.1 Review of project

Stated simply, the purpose of the project was to construct a network incorporating novel design features and potentially capable of carrying a wide range of communications traffic. Having built the network, a variety of experimental systems were attached in order to test it and assess how far it could meet its objectives. Many of the results of these experiments can already be found in the literature (see Bibliography) and all have been covered in earlier chapters here. Imposed on the project as an afterthought was the not insubstantial activity of mounting the public demonstrations described in Chapter 10. Though expensive of effort, these demonstrations were never among the primary purposes of the project and no more will be said about them here.

11.1.1 *Architecture*

Central to the project was the evolution of the UNIVERSE network architecture, based on two fundamental ideas. The first of these encompassed the notion of a set of sites, each served by a packet network composed of one or more interlinked LANs, and with the sites connected to each other by a high-speed wide area network in which only a very small number of hops is needed between any pair of sites, ideally a single hop. The second fundamental concept was that of the LVC; this was born of a desire to carry on a packet network those services usually conveyed by circuits, and is an approach offering possibilities more promising for voice and image traffic than conventional schemes using ARQ and flow control. The first of these ideas is now generally accepted as an important method for constructing networks; the idea of the LVC, however, has so far attracted relatively little attention, although this situation is changing with the growing interest in fully integrated networks.

Use of LVCs implies a shift in responsibility for the operation of the network to the end-points with respect to the control of congestion and provision of reliability within the network. Congestion within a network, where the network is forced to discard data, can only be prevented by limiting the flow of data into the network. In circuit-switched networks, this is achieved by two mechanisms: limiting the bandwidth associated with each circuit, and limiting the number of circuits in use at any one time (call blocking). This contrasts with the traditional mechanism used in packet networks of increasing the packet transit delay, a mechanism clearly unsuitable for time-critical traffic. Since neither call blocking nor bandwidth limitation was implemented in UNIVERSE, the only recourse was for the network switching elements (bridges) to discard packets.

Bandwidth limitation on LVCs may be achieved by adopting rate-driven methods of communication between end-users. These are also particularly appropriate for networks with long end-to-end delay, such as those with satellite segments. Use of a rate negotiated between the two end-points of a connection also implies the need for bandwidth reservation in the network switching elements. This is essential for voice communication and is also necessary for some categories of video communication. Partial reservation, coupled with a range of transport priorities, may also have application in meeting specified performance limits for remote access to transaction systems.

An aspect which LVC- and datagram-based networks have in common is in the provision of reliable, sequenced communication. In both cases this must be provided on an end-to-end basis, though the LVC does have the advantage of not introducing reordering of packets.

The UNIVERSE LVC had asymmetric initial connection: the OPEN was out-of-band but the OPENACK was in-band. This arose from the requirement to maintain compatibility with single ring operation and from the goal of keeping the delay associated with initial connection (and SSP communication) to a minimum by enabling reply data to follow an OPENACK without pause. This was especially important in view of the delay imposed by the satellite. Taking both OPEN and OPENACK out-of-band would simplify and improve the performance of bridge code for the data phase; this would be particularly important, if further control messages such as closure were to be implemented, to reduce the cost of processing every packet on an LVC. However, use of completely out-of-band initial connection does have the disadvantage that the two ends are not immediately synchronized since the recipient does not know when the initiator will receive the OPENACK, that is the instant when the LVC is fully established, and cannot transmit data without further notification because the data may overtake the OPENACK. Completing end-to-end synchronization in general causes a further round-trip delay; this extra delay is in practice only incurred in the relatively unusual circumstance that the call recipient is the first to send data on the LVC.

As implemented, the network had no explicit close. It was argued that as time-out mechanisms were necessary for 'long-stop' protection of network resources from malfunctioning hosts, all connections could be left to time out. This seemed quite workable since the only resources being consumed were port mappings in bridges. Issues related to charging were not addressed by the architecture, which was mostly designed to be the minimum necessary to enable the network to work well enough to allow experimentation. Bandwidth reservation, which obviously consumes valuable resources, and charging are examples of features which would make the requirement for explicit closure of LVCs imperative.

The SSP, provided as one of the primitive network access mechanisms, was a success for distributed systems activities and was used extensively. It enabled the direct translation of local systems to wide area systems with little or no change. However, especially in the absence of explicit closure, it is very doubtful if it was worth distinguishing it from the LVC initial connection protocol: the difference in bridge processing was minimal and probably outweighed by the effort of maintaining the distinction. If the LVC were used instead of SSP for short interactions across the network, it would be important to keep the delay associated with call initiation as short as possible. In the case of completely out-of-band initial connection, if the reply to a transaction request were sufficiently short, it might be combined with the OPENACK to reduce the delay to that of the asymmetric option used in UNIVERSE.

Time-outs were a problem, particularly as they had end-to-end significance and insufficient information was available to enable a proper choice of their size. Problems with their use in connection with SSP were very much exacerbated in the case of services which used nested SSP(s). A typical example was in finding a service and setting up a route to a remote service, as part of which a nameserver might have to perform a lengthy interaction with a remote nameserver. As a consequence, a relatively long time-out was required; for continual use, cacheing was essential to retain any semblance of performance, especially in the presence of errors. An unpleasant consequence of using time-outs dictated by worst-case requirements is that local performance is substantially degraded in the presence of errors. A further problem is that since SSPs leave state in bridges which is only removed on expiry of a worst-case time-out, any substantial use of the mechanism will consume considerable amounts of port number space and memory in bridges. In consequence, the SSP is unsuitable for widespread use in transactions with significant delay.

A successful aspect of the architecture was its ability to serve as the basis for conventional ISO layered protocol implementations and for experiments with novel protocols. While the latter ability was exploited in several experiments, particularly for voice, an omission was the lack of any serious work on the design and implementation of an appropriate transport protocol engineered to take account of the satellite delay. BSP and its derivatives operated reliably enough but were restricted by the window of one and the satellite delay to a fraction of a per cent of the available bandwidth. In the context of bulk data transfer, this problem received considerable attention in the contemporary NADIR project (Azouz *et al.*, 1983) and more recently in the NETBLT experiments (Section 9.7 and Clark *et al.*, 1987). Although no ISO protocols were implemented in UNIVERSE, interworking with JANET was achieved using transport-level relaying; while JANET does not use ISO protocols, nevertheless its general structure and layering accord with ISO principles.

A novel and interesting feature of the architecture was the incorporation of a broadcast service. Though broadcast is intrinsic as a mechanism in satellite networks, it is uncommon as a service offered by a packet satellite network. It is more typically associated with simplex satellite networks, such as those used for television broadcasting and, more recently, for disseminating a wide variety of digital data. As part of a duplex packet satellite network it is in need of further exposure to enable a more realistic assessment to be made of its merits (or otherwise) and appropriate areas of application to be identified. Quasi-continuous mode applications, such as status monitoring, seem natural. File distribution appears possible but is evidently not simple and requires special care for error recovery and some attention to initial coordination ('rendez-vous'). In the project, a basic experimental approach was adopted, from which it was concluded that the use of broadcast had some potential as a means of optimizing overall satellite usage when some of the traffic was multicast file distribution. To realize this potential it would be necessary to develop a properly robust protocol, capable of adjusting to the presence of congestion and withstanding loss of data due to corruption. Work in this area was undertaken in the NADIR project (Azouz *et al.*, 1984) and has also been the subject of study by Gopal and Jaffe (1984), Sabnani and Schwartz (1985), and Daka and Waters (1988).

The primary purpose of nameservers is to relieve software and hardware components from dependencies on known fixed addresses for hosts and services. Although naming structure and management were not main areas of study, experience with the use of nameservers in the project supported this general principle. The manner of their use in the project also provided the added benefit of devolved management for network names. However, the mechanics of name-table management were manual and, even with the necessary privilege, a manager could not alter name tables in any uniform, organized fashion, still less perform the operation remotely. In essence, the problem is a specialized instance of distributed database management, where the requirement is for frequent and rapid read access to a relatively static set of tables. Update is quite infrequent but attention to maintaining consistency and update authority are essential. Although the absence of most of these features did not prevent either the operation of the network or the execution of experiments during the project, the same would probably not be true for any larger project and solutions would certainly be necessary in a service environment. Another unsatisfactory aspect of the network implementation was the incorporation of routing into the name-serving function. Though apparently another database management problem, the principles and patterns of use make the dynamics of routing-table management quite different from those of name-table management.

Routing was fixed in the UNIVERSE network implementation. This was originally conceived as an initial simplification but, as no further work was undertaken in this area within the project, it was in effect built into the architecture. Limited routing information was loaded into the nameservers, whence it was read by the bridges when they started up. For any destination domain or subnet, a unique bridge was defined which constituted the next 'hop' towards the desired destination. Thus the network effectively ignored the presence of loops in its topology and for every pair of subnets defined a single, fixed route between them. There was no ability to exploit such loops, whether for automatic load balancing among bridges or for circumventing failed network segments; nor was there any ability to alter routing manually, other than by the laborious and impractical recourse of reloading all relevant nameservers. This was sometimes troublesome in the project, particularly at UCL where the only loop existed. Apart from the obvious operational inadequacies, it was a serious hindrance to some diagnostic and monitoring activities.

In considering the scaling properties of the architecture for use by a substantially larger community, a number of issues take on added significance. Devolution of name management appears to be essential, partly to spread the load of the task and allow timely response to necessary changes and partly because, for networks whose operation is shared among a number of organizations, it is not practical to separate this aspect from the other operational responsibilities. A fundamental parameter of the network which does not scale is the number of satellite hops. If geographical coverage or any other consideration caused this to increase, application of the network would be biased further in favour of those services in which bulk data transfer predominates and delay is tolerable. The scaling properties of the individual (LAN and satellite) segments of the network are reviewed below. However, it may be observed that the number of nodes attached to a single service LAN is typically about 100 and seldom exceeds 200. This is as much a consequence of organizational and managerial convenience as a reflection on the shortcomings of current LAN technology. Thus it may be expected that, while the number of subscribers served by a single LAN will increase in the future, equally important will be the requirement to increase the capacity and number of interconnections between LANs. In terms of the basic concepts of the UNIVERSE network architecture there seems no fundamental obstacle to increasing the number of bridges in a network. Of course, this would highlight the inadequacies of routing and name table management already noted above. Increased traffic between LANs and the presence of a large number of bridges in a data path would focus attention on the throughput and delay characteristics of bridges. In UNIVERSE, delay was dominated by the intrinsic satellite delay, so that throughput was the major concern. Ultimately, bridge throughput was a problem since, of the bandwidth of approximately 1 Mbit/s potentially available on a point-to-point basis on either the satellite or LAN segments, a maximum of about a quarter of this could be delivered by the bridges. Though not in practice a problem for individual data streams in UNIVERSE, this was a problem when bridges were required to handle a number of data streams. Clearly, in any larger network this would be a major issue.

An important area vital to the operational viability of any network is network management. Although GEC and UCL in particular considered a number of issues relating to this topic, the subject was one which received scant attention in the project as a whole. For the conduct of the project this was not too serious, as the community of researchers was small. It is the common experience, confirmed by this project, that network management tools such as packet tracing, statistics gathering, remote diagnostics, etc. are hard to provide as afterthoughts. For any larger project or in a service operation they must be incorporated at the initial design stage. The problem in so doing in a high-speed network is that many of the components providing these tools must also provide high throughput and low delay for data transport.

11.1.2 *LANs*

Contemporary with the project, LAN development underwent an intense period of standardization, conducted under the auspices of the IEEE 802 committees. The result was not a single standard but an effective acceptance of the *de facto* situation of there being three major LAN technologies (ethernet, token bus, and token ring), all of which had been developed in the 1970s (IEEE, 1984, 1985*a*, 1985*b*, and 1985*c*; Tanenbaum, 1988). The Cambridge slotted ring, based on ideas conceived at the beginning of the decade, was developed somewhat later than the ethernet, though early enough to exert some influence on aspects of token ring design. The slotted ring is an essentially simple scheme, certainly simpler than the token ring. The shortness of the minipacket also implies a modest requirement for fast logic in any implementation, since it is only required to cope with very short bursts of data occurring at the full rate of the basic transmission medium. It was unfortunate that in spite of these apparent implementation advantages over its rivals, early attempts at production of a VLSI version of the Cambridge ring failed. This, coupled with its having attained international standardization (ISO, 1988) only quite recently, has resulted in this technology having achieved a negligible share of the market, even in the UK.

While the shortness of the minipacket was a potential advantage for low-level hardware implementations, it was also a weakness because it was responsible for a disproportionately small amount of the basic transmission bandwidth being available for data. Furthermore, it was a barrier to the production of an interface which could multiplex reception and transmission of basic blocks at the minipacket level: this is essential to prevent access to multiaccess services being locked out by slow clients. The difficulty with such an interface is in coping with the high rate of interrupts caused by the short minipacket length.

For the project, the Cambridge ring was satisfactory: the network bandwidth was not limited by its use and the architecture was not dependent upon it. Interfaces and all the expertise were available in the UK. The same could not be said of any other comparable LAN available at the time and the project would probably have been generally more hampered by the use of another LAN. Nevertheless, the lack of a multiplexing interface was a problem, particularly for multiaccess network elements such as bridges.

Early use of VLSI chips for the Cambridge ring would have corrected the excessive shortness of the minipacket by extending it to a maximum of eight bytes. This would at once have made easier the task of developing a multiplexing interface and increased the point-to-point bandwidth, while preserving an important advantage of the Cambridge ring, its suitability for voice (low latency and guaranteed access to the transmission medium).

11.1.3 *Satellite TDMA system*

Turning to the satellite segment of the network, this was conservative in overall design. Hardware work was concentrated on replication, installation, and commissioning, with just sufficient effort devoted to making necessary adaptations to existing equipment; development of the software required to implement the architecture was the major concern. The inadequacies of the resulting implementation were anticipated at the beginning of the project and were accepted on the grounds of expediency to enable construction of the network in the time available. At the same time, a parallel activity was initiated to design a TDMA controller with more appropriate performance specifications.

The UNIVERSE 2 Mbit/s channel was one of a number provided by FDMA use of the bandwidth provided by a single transponder on OTS. The conservatism of the UNIVERSE TDMA design, exemplified by the long time frame, large guard times, and high transmit power, led to substantially trouble-free operation of the TDMA system. The exception to this occurred during the Telecom '83 exhibition at Geneva when all the FDMA channels were occupied simultaneously and UNIVERSE was required to reduce its transmit power, whereupon operational problems with the TDMA scheme were experienced.

However, although experimentation was possible, much higher performance was needed from the controller. This was a limiting factor in the project, both in respect of rather crude use of a large time frame and the lack of guaranteed access for time-critical traffic such as voice. From an operational standpoint, the main problems were due to earth-station unreliability. This probably stemmed from the fact the main requirement in Europe has been for small numbers of high-powered, attended stations, using duplicated equipment for reliability, and there has been only a very small market for highly-replicated, low-cost stations.

Associated with the TDMA scheme used in the project are a number of scaling problems. Allocating each station a permanent small slot in each time-frame, so that stations not currently transmitting can make requests for bandwidth allocation, can only be used for networks with a small number of stations. This limitation can be removed by allocating a fixed small number of these slots in each frame and then either allowing stations to contend for their use or allocating use cyclically to stations. For example, using the new access equipment in the SATINE-2 experiment (Hine *et al.*, 1988), four small slots are included in each TDMA frame, and these are allocated cyclically to a maximum of 64 stations. With a shorter time-frame, the maximum wait for a slot is limited to half a second. The second scaling limitation is the total bandwidth available to the TDMA scheme. The equipment used for UNIVERSE was limited to 2 Mbit/s, though only 1 Mbit/s was available for data because of the fixed, half-rate forward error correction encoding; the new access controller is designed to support data rates up to 8 Mbit/s when no encoding is used. At the present time, above this speed a regime is entered where increases in performance are likely to incur disproportionate increases in costs. Currently, 8 Mbit/s would probably be adequate for a private network catering for a few tens of sites on a continental scale and use of FDMA would allow a number of these to operate through a single transponder. A station able to receive and transmit on two or more frequencies could act as a bridge between such networks, at the expense of the delay associated with two satellite hops.

11.1.4 *Distributed systems and multi-media communication*

In the creation of end-user systems two stages can be distinguished. First of all, underlying techniques and individual components are developed, and then comes the assembly, coupling, and customizing of these components to form an application system. In broad terms, distributed systems development can be considered to have entered the second stage, whereas multi-media applications are still very much in the formative first stage.

The distributed systems experiments and demonstrations showed that the network architecture was certainly capable of supporting the wider distribution of several elements of the Cambridge distributed computing system (authentication and remote fileserver access are examples). Where remote data access was required, it was demonstrated that cacheing could be used quite effectively to counteract some of the loss in performance due to network delay. Furthermore, there were promising signs that wide area distributed applications such as USE could be constructed. Had the project continued, the initial results of this experiment were sufficiently encouraging to have ensured its further development to the stage where it could have undergone wider trials. A particularly significant and interesting feature of this application was that wide area distribution was intrinsic: the assumptions were that specialized resources existed in different systems, that both these and the target system(s) were widely dispersed, and that bringing them together was either costly or impossible. Stated thus, it appears evident that USE could be regarded as typical of a large class of future applications.

Intrinsic to the majority of multi-media communications applications is the transport of data under a variety of real-time constraints. These criteria impose quite different requirements upon the communication and switching fabric from those arising from the fast, transaction-style traffic needed to support conventional distributed computing systems. These contrasting requirements were thrown into sharp relief by the exploratory attempts made in the project to mount some simple distributed multi-media communications applications. The need for bandwidth reservation in the network has already been mentioned, particularly in connection with voice. As the network load increases, so it may be appropriate to allow rates of compression to vary. The requirements for recording and playback communication are substantially less stringent than those for conversation. While there is a great deal of work to be done in these areas to understand packet voice better, the capacity and access parameters of the network were good enough to enable a start to be made in studying these problems. The situation for video, on the other hand, is quite different: to reach a similar position in even the most basic network parameters of capacity and end-to-end throughput requires an improvement of one or two orders of magnitude.

11.1.5 *Organization*

Details of the way in which the project was managed have been described elsewhere (Chapter 1 and Appendix A), from which it will be evident that the management arrangements were, on the whole, relatively simple and lightweight. This was a considerable boon to the progress of the project. The unsatisfactory feature of having a steering committee with financial control but distanced from the technical aspects of the project, coupled with a principals' committee with responsibility for execution of the project but without full executive control has already been remarked upon: suffice it to say that efforts should be made to avoid this mistake in the future. A respect in which the project was lucky was in being relatively free from conflicts of interest. This is a very difficult area: the origins of such conflicts are not always evident but it is essential to success that the detailed goals of each partner in such an endeavour can be accommodated within the project.

The cost of collaboration is high. UNIVERSE was fortunate in that all the partners were situated in the South East of England, within easy reach of London for regular meetings and of each other for site visits. Nevertheless, the amount of time and effort that must be allocated in the cause of collaboration is very high. In this respect, the availablility of JANET was a valuable aid in the day-to-day activities of such collaboration. In terms of both the mechanics of collaboration and the risk of differences in perception of objectives, it is the authors' view that a larger set of partners would have led to a much looser federation, substantially increased collaboration costs, and increased risk of being ineffective.

Lack of common infrastructure for software development resulted in the use of a wide range of hardware, which in turn led to heavy consumption of effort in interfacing. With hindsight, it might have been better to have made some sweeping decisions on hardware and software to be used in the project and to have accepted the initial pain (and expense) of coming to terms with this at each site. Nevertheless, a degree of commonality was achieved for all the network and many of the infrastructure components, so that in many cases there was in practice a single investment in software for the whole project. This, together with key technical decisions, such as basing the network architecture and satellite protocols on previous work, and the availability of JANET to control satellite network testing and to exchange software were all crucial to achieving results in the requisite timescales. In the event, the project was probably the first to put into place an integrated local and wide area system which covered different organizations and used such a high degree of common infrastructure.

11.2 Related developments

11.2.1 *Architecture*

As might be expected, the problem of interconnecting LANs within and between sites has received considerable attention. For a given type of LAN, such as ethernet, various forms of repeater are available for interconnections within a site. Such schemes operate at the physical level and have severe limitations of range and absence of filtering which debar their use outside a single site. At the lowest level for wide area interconnection, schemes using bridges that operate at the MAC level have been devised (Berntsen *et al.*, 1985; Tanenbaum, 1988). These vary from simple bridges that use static tables for routing packets between LANs, through schemes that use source routing, to bridges that use very little a priori information and learn the topology of the internetwork by observing traffic. The latter are of some interest not only because products are available but also because the original work is contemporary with UNIVERSE and represents another approach to constructing a wide area network which is like a giant LAN. These ideas have been developed particularly by Digital Equipment Corporation and Vitalink Communications Corporation in the USA (Hawe, *et al.*, 1984; Hart, 1985). The basic principles for a bridge connecting two ethernets can be summarized quite simply. The bridge operates in promiscuous reception mode on both sides. By observing traffic and noting the origin of every packet on each side, the bridge builds up tables of station addresses observed to be on each side. The algorithm for when to forward a packet is to do so if its destination is not known to be on the same side as its origin. This algorithm assumes that all traffic is two way, so that packets will eventually be observed originating from an address to which the bridge has previously had to forward in ignorance of its whereabouts. Obvious prerequisites for the mechanism are a global addressing scheme for the collection of LANs and a promiscuous reception mode. In a topology with potential loops, the bridges detect these and the first to do so for a particular loop reverts to listening only, to reduce the topology to a rootless tree. A more advanced version introduces active probing of bridges by each other to determine the optimal tree which spans the mesh of LANs (Backes, 1988; Hart, 1988; Sinkoskie and Cotton, 1988). The use of broadcast in such an arrangement presents problems to which no satisfactory solutions are readily available. Because of the uniform addressing at the MAC level and the availability of promiscuous reception mode for the three IEEE 802 standard LANs (ethernet, token ring, and token bus), this type of bridge can operate between any two of the IEEE LANs and products are available for bridging among all three.

For interconnection within a site, bridges are often used in conjunction with a backbone LAN. Such a backbone does not have end-user hosts connected but is used exclusively to link LANs together. If MAC-level interconnection is used, then all the stations on a backbone are bridges. For wide area connection, bridges may be joined by leased lines or by satellite links. Of particular interest is the TransLAN range of products from Vitalink (Hart, 1985 and 1988) for use with satellite links. Each site transmits on an assigned frequency and receives on a number of frequencies, one for each of the other sites. The TransLAN bridge at each site is in effect a set of bridges, each of which handles the bridging to one of the remote sites. Other configurations, in which multiple satellite hops are admitted, are also possible, to reduce overall cost by allowing some locations to use low-power, smaller earth stations. A particularly common form of this is a star topology in which the hub site can communicate with every site but other sites can only communicate with the hub and must communicate with each other by two hops via the hub. The potential advantage of bridging at the MAC layer is that, apart from the delay and its consequences (such as the necessity of adjusting time-outs), the host systems can communicate in the same way as if they were all connected to a single LAN. In this sense the objectives are similar to those of project UNIVERSE.

Apart from MAC-layer bridging, the other forms of general-purpose interconnection of packet networks consist of network and transport level relaying, typically based on internet datagram routers or transport gateways. Relaying at these three levels is now primarily a subject of study and debate within the standards organizations. Important though this is, in the context of integrated services it has to be remarked that neither of the conventional approaches is currently suitable for time-critical traffic requiring short latency and delay combined with low jitter. For connectionless (datagram) techniques (MAC-layer bridges and datagram routers), call blocking is not possible and recourse has to be made to use of time-to-live and priority applied to individual datagrams. In the absence of any absolute way of preventing further traffic entering the internet, these palliatives cannot provide the necessary guarantees on performance in terms of those network parameters of importance to real-time traffic. For connection-oriented relays (gateways), it would be necessary not only to discard the usual mechanisms of flow control and ARQ error recovery but also to provide call blocking and bandwidth reservation facilities. At present, these are not characteristics of standard gateways.

Recently, there has been an upsurge of interest in the possibility of using Integrated Services Digital Network (ISDN) technology for the interconnection of LANs over a wide area. ISDN technology is being used to convert analogue telephone networks to digital operation. The basic unit in this technology is a synchronous 64 kbit/s duplex channel, called a B-channel, that is used to carry a PCM voice call. Signalling in ISDN networks is out-of-band and is carried on D-channels operating at either 16 kbit/s or 64 kbit/s. The signalling protocols are packet-based using X.25-like protocols for the link and network layers. Customer access is either at the basic rate, which offers two B-channels and a D-channel, or at primary rate with 30 B-channels and a D-channel. A number of types of H-channel, operating at higher-rate multiples of 64 kbit/s, have also been defined but not generally implemented. An overview of ISDN can be found in Tanenbaum (1988); Rutkowski (1985) contains details of the standardization process for ISDN which has taken place over a number of years.

B-channels or H-channels could be used as dial-up point-to-point links between LANs. However, in the 1988 CCITT recommendations (CCITT, 1988), a framework for services has been included which will provide a data transfer procedure, called frame relaying, in which only the lower part of the link level will be handled in the network. Flow control and error recovery for these services will operate on an end-to-end basis. Call setup will be out-of-band using the D-channel in a same way as for B-channel setup.

Further into the future, a favoured technology for implementing broadband ISDN (B-ISDN, also called Integrated Broadband Communication or IBC), an ISDN offering services at rates above 2 Mbit/s, is Asynchronous Transfer Mode (ATM) (Littlewood *et al.*, 1987; Fisher *et al.*, 1988). (*Note* ATM is a generic term for a number of packet-based transfer modes under consideration; strictly, the scheme outlined here is one such called Asynchronous Time Division or ATD.) In this scheme information is transported across a network in fixed length blocks called cells, consisting of a header field and a data field. The header identifies the virtual circuit to which the cell belongs. Cell sizes of 32 or 64 bytes are under consideration. Time division multiplexing of cells is performed asynchronously in the sense that the intervals between successive cells on a virtual circuit are variable and determined by the source rather than the network. The obvious similarity between the small, fixed-length cells of ATM and the structure of slotted LANs such as the Cambridge fast ring (see next section) will be apparent.

11.2.2 *LANs*

During the course of the project the deployment of LANs as operational, site-wide computer networks started. Subsequently, with the widespread introduction of personal computers and customized workstations, LANs have become the standard means of providing local networking. However, these LANs are not generally used to carry multi-media traffic. Speeds in the 10 Mbit/s region are not adequate to carry substantial numbers of voice channels mixed in with computer data and are well below requirements for video. Higher speeds are also required for some computing applications, particularly those involving high-resolution graphics.

To meet these requirements, development work on LANs providing an order of magnitude increase in speed has been undertaken. Work in the USA has concentrated on extending token ring technology using fibre optic transmission. This has led to the definition by the American National Standards Institute (ANSI) of standards for a Fiber Distributed Data Interface (FDDI) (Stallings, 1987). Ross (1986) gives an introductory review of this work and Tanenbaum (1988) contains an overview. These standards define the physical layer, the MAC layer, and the management functions for a 100 Mbit/s ring LAN using token-passing access control. The media access control scheme is similar to that used in the existing IEEE 802.5 token ring standard, but the physical layer is quite different. A chip set to implement FDDI is being developed by Advanced Micro Devices Inc. in the USA, but is not yet generally available at the time of writing.

In the move to higher speed LAN technologies, the slotted ring scheme has again come under consideration. A version of the Cambridge ring, called the Cambridge fast ring (CFR), has been defined which is suitable for operation at speeds of the order of 100 Mbit/s (Hopper *et al.*, 1986). In the CFR, the data part of the minipacket has been increased to 32 bytes, and 16-bit rather than 8-bit addresses are used. A chip set is available that implements the CFR definition, running at about 50 Mbit/s. LAN equipment using this chip set is currently the only fast LAN equipment that can be obtained at a cost which is feasible for use with personal computers and workstations. A revision of the slotted ring technique, called the Orwell ring (Adams and Falconer, 1984), has been developed in the UK by BT. This design pays particular attention to the problem of carrying a large volume of time-critical traffic on a LAN, such as would be imposed by multimedia applications or its use as the basis of a distributed private branch exchange (PBX). A feature of the design is that instead of a data slot being emptied by the transmitting station, as in the Cambridge designs, it is emptied by the receiving station, thus increasing the overall throughput of the ring. A chip set for this scheme is also under development.

A second thrust in LAN development has been to extend the size of the physical area over which single LANs can operate, a move from the site LAN to the town LAN. The 802.6 committee of the IEEE has been charged with developing a standard in this area under the title of the *metropolitan area network*, or MAN, with the specific requirement to be able to carry voice and other guaranteed-bandwidth services. The currently favoured technology for MAN standardization is the scheme developed in Australia known initially as Queued Packet and Synchronous Switch (QPSX), now generally called Distributed Queue Dual Bus (DQDB) (Newman *et al.*, 1988).

However, work is at quite an early stage and it is not clear just what role LAN-derived technology will play on an urban scale.

11.2.3 *Satellite*

Two other satellite-based computer networking projects were contemporary with UNIVERSE, the Wideband Satellite project, sponsored by DARPA in the USA (Edmond *et al.*, 1986; Heggestad and Weinstein, 1983), and the NADIR project in France (Huitema and Radureau, 1982; Meunier *et al.*, 1983; Valet, 1983). The Wideband project concentrated on the development of sophisticated hardware and software for satellite access with the aim of making efficient use of the satellite channel for carrying a mixture of telephone traffic and computer data. As has been mentioned in Chapter 4, the work in project UNIVERSE on a new satellite access controller was influenced by early work in this project. After an extended period of technical development on the access control equipment (which incidentally served to confirm the UNIVERSE decision to proceed with very simple equipment) the Wideband Satellite project proceeded to a period of experimentation (Heggestad, 1984; see also the foregoing references and Section 9.7). By contrast, in project NADIR, hardware simulation was used in place of a real satellite, thus obviating problems of satellite access. Work here concentrated on producing protocols that optimize the use of fixed bandwidth point-to-point and point-to-multipoint channels for data transfer (Azouz *et al.*, 1983 and 1984).

Developments in Europe in the use of satellite links for computer networking since the end of the project have been rather disappointing. Operational versions of the OTS, called the European Communications Satellite (ECS), have been brought into service. However, these satellites are mostly used for television relay. Point-to-point and point-to-multipoint channels are available for lease but are expensive and little used. A similar series of satellites, called Telecom 1, have been launched by France. These have footprints tailored for communication within France and neighbouring regions. Again the use for computer networking has been limited.

Following the end of UNIVERSE, the atmospheric science community in the USA was sufficiently encouraged by what had been achieved in the project to consider a combination of LAN and satellite technologies for the basis of a new communications infrastructure then being planned as part of a larger effort to support academic research in atmospheric science. This programme, known as Unidata (University Data) (Cooper, 1987), chose to separate data distribution from general purpose site interconnection and supercomputer access. A C-band simplex broadcast satellite channel is being used to provide delivery of a special Unidata data stream as well as the US National Weather Service data to a community of some 70 institutions.

Data from the broadcast stream are received, selected, and stored by a workstation-based data management system connected to a LAN and are accessed by other local workstations and personal computers via the LAN. Access to supercomputers and interconnection of sites is being developed in conjunction with NSFnet, the US academic research network backbone sponsored by the National Science Foundation (Jennings *et al.*, 1986). The beginnings of NSFnet coincided with the early stages of Unidata and it was under the auspices of the former that nine sites (about half of them Unidata sites) are using Vitalink Ku-band earth stations and TransLAN equipment (Section 11.2.1) to provide a high-capacity, bulk data subnet within NSFnet.

11.2.4 *Distributed systems*

The principles of distributed computing system design described in earlier chapters were already well known before the project began. During and since the project, products have been developed which incorporate such principles and it is has been possible for several years now to purchase a great variety of equipment which can be configured to provide customized distributed systems based on the current generation of 10 Mbit/s LANs. Typical of such equipment is the Sun workstation range, which features diskless workstations supported by one or more fileservers. The Sun fileserver interface specification, known as Network File System (NFS), has been made public in an effort to encourage standardization in this area. Other, similar components, such as the Novell fileserver, are available to provide shared, remote storage management for groups of IBM PCs (or equivalents). For detailed treatment of many of the significant research and production systems developed over the last decade or so, the reader is referred to Coulouris and Dollimore (1988).

One particular technique which has matured during the first half of the decade is the use of remote procedure call for interactions among distributed system components. Apart from encouraging the analysis and design of systems so that there is a clear, explicit separation of program flow into a set of sequential threads which execute asynchronously (possibly in parallel), the RPC also embodies sufficient mechanism to enable nested interactions to take place among a set of system components without the need to re-engineer end-to-end time-outs, thus providing a solution to the problems, described in previous sections, of using nested SSPs.

The subject of multi-media systems is as yet in its infancy. The shift from analog to digital technology for handling video and voice, which promises the possibility of a wide range of novel applications in the future, is accompanied by the need for massive increases in capacity in all the traditional resources associated with computer processing: memory, backing storage, processing cycles, and communications bandwidth. There are now signs not only of the emergence of technologies to meet these new demands but also of integration beginning to take place. One such example is the recent announcement by NeXT Inc. in the USA of a workstation referred to as 'the cube' (Thompson and Baran, 1988). Apart from the use of optical backing storage, the system includes a signal processing chip for application use and integral facilities for sound input and output. The introduction of this sort of desk-top facility at a reasonable price (expected to be around $6500 in the first instance) and its combination in future with higher capacity, integrated communications will doubtless open up a new range of distributed systems applications and research topics.

11.3 Postscript

During 1983 the partners worked on a proposal for a continuation project, UNIVERSE II. By this time BT had announced the Megastream service of point-to-point leased lines operating at 2 Mbit/s and it was proposed that the industrial partners would build and operate a network based on the use of these lines with LANs acting as switches to implement the UNIVERSE network architecture. This network was to be used for the continuation of a programme of application system experiments. As a research activity, work on integrating this network with a satellite network was to continue using the new access controller (Chapter 4) working to one of the European Communications Satellites (ECS-2). This proposal was presented to the newly formed Alvey Directorate of the DTI at the beginning of 1984. After a rather frustrating period of uncertainty, it became clear that the Directorate did not wish to fund a large integrated project in computer networking.

Instead the Directorate decided to proceed by commissioning the construction of a high-speed packet network using some existing technology and calling for proposals to use this network for applications experiments. Logica were asked to review technologies that could be bought in and to make a proposal. Predictably, it was found that there was no equipment that could be used to construct a packet network without incurring extensive development. However, quite unexpectedly they concluded that a small high-speed circuit-switched network could be constructed from existing equipment and that with some modest development work this could be made to conform to ISDN standards. The Alvey Directorate accepted this advice and towards the end of 1984 commissioned BT with Logica to build a small pilot ISDN to be called the Alvey High Speed Network (AHSN). The UNIVERSE partners together with two new partners, Acorn Computers Limited and the University of London Computer Centre (ULCC), formed two groups and submitted two proposals (called projects Admiral and Unison) to exploit the AHSN. These were accepted by the Directorate and started early in 1985. The project Admiral consortium, consisting of BTRL, GEC, UCL, and ULCC, concentrated on issues of networking technology, particularly network management and control (Pope *et al.*, 1986). The Unison consortium of Acorn Computers Ltd, Cambridge University, Logica Ltd, LUT, and RAL has worked in the area of communications and basic components for office multi-media applications (Tennenhouse *et al.*, 1987). All the partners in both projects were supplied with primary rate (2 Mbit/s) access to the AHSN. At the time of writing, project Admiral has recently been completed and Unison has entered its last year of activity.

The work on a new satellite access controller started in the project has been continued in a European partnership between GEC and RAL in the UK, CNUCE at Pisa in Italy, and the Graz University of Technology in Austria, working under the European Commission's COST 11 programme. The new satellite access controller described in Chapter 4 has been completed. Earth station installations in the three countries working with the ECS-2 satellite have tested its capabilities. The results of these tests were reported at the 1988 EUTECO conference held in Vienna (Hine *et al.*, 1988) and operation of the controller was also successfully demonstrated there. This style of 'slim' TDMA working has had little influence as yet on the communications services offered by common carriers. However, in Europe there are now some signs of a growing interest in data communication via satellite and in the use of slim TDMA for this purpose.

From the technological point of view, the prospects for high-speed computer networking appear promising. ISDNs offering connections at basic and primary rates are coming into place in Europe, North America, and Japan, in spite of the many problems which beset its introduction (see Tanenbaum, 1988). There is a realization that this development, large and important though it is, is only a step towards IBC. It is anticipated that IBC will exploit the enormous digital transmission capacity that can be obtained from fibre optic cables to make carrier services above 2 Mbit/s available for the support of all communications services, including high-definition video. In Europe, research and development for IBC is co-ordinated in the RACE (Research in Advanced Communications in Europe) programme of the EEC. This programme is targeted for deployment to begin in the late nineties. The development of multi-media, end-user equipment will certainly keep pace with the deployment of higher speed communication services. Personal computers with voice and video capabilities are on the verge of appearing at costs appropriate for widespread use in the research and development of applications. Low-cost compression schemes will enable such personal workstations to handle high-quality still video frames and limited moving video images integrated with windowed display systems. Auxiliary components such as cameras, scanners, and laser printers are also generally available in suitably affordable forms. Development activities in the immediate future should be able to produce a wealth of integrated applications from these basic facilities.

Notwithstanding the optimistic prognosis for the technology, there are numerous organizational and economic obstacles to be surmounted before any form of IBC can become ubiquitous. Confidence in the general levels of integrity and security of such all-pervasive technology may well become a key issue for both business and domestic acceptance. Particularly in the case of security, there are issues of public policy and of technology, both of which will take time to resolve. Economically, the deployment of new ground-based plant covering a wide geographical area is an inherently slow process, involving very large capital investment. Traditionally, such investment could be expected on the basis of the perceived market. Unfortunately, it is too early in the application development cycle for it to be evident yet where the significant new markets may be which have sufficient potential to underwrite the required investment. In the past, the argument has been that telephone use will always be dominant and thus networks must be optimized to handle voice. That there are other substantial markets can be seen, for example, from the rapid rise in the use of the telephone network for automated credit verification and for facsimile. The first of these is an example of rapid, transaction-style traffic, likely to see explosive growth if plans for electronic funds transfer at point of sale and similar developments come to fruition. In terms of sheer volume of data transferred, it seems evident that voice traffic would no longer dominate if image-based communication achieved even a fraction of the IBC market. Another avenue of speculation suggests that the use of technology such as IBC may ultimately be forced by a combination of the failure of older, traditional communications technologies and by changing work practices brought about by apparently unrelated circumstances. As an example of the former, declining levels of conventional mail service in many places have already been responsible for substantial growth in both facsimile and courier services. Moreover, considerable growth is also taking place in industries for which it is only necessary or appropriate for people to commute to a conventional office or other centralized place of work for a fraction of the time. One may speculate that such changes, coupled with the ubiquitous problems of urban transport congestion, may reach the stage when heavy investment in IBC technologies will be attractive both economically and socially.

Appendix A. Committees and Working Parties

(The name of the chairman is shown for each committee and working party.)

Committees

Steering Committee	*J H H Merriman*
Principal Investigators' Committee	*P F Linington*
SERC Management Committee	*G Manning*
Exhibition Steering Group	*D B Thomas*

Working Parties

Protocols and Implementation	*C J Adams*
Construction and Operations	*A D Marshall*
X.25	*S R Wilbur*
Distributed Computing	*R M Needham*
Security and Encryption	*D King*
Voice and Images	*J W R Griffiths*
Demonstrations	*P T Kirstein*
Cables	*P T Kirstein*
Teletex	*W R Tuck*

Appendix B. Project Reports

(These reports are available from either of the authors or the library at Rutherford Appleton Laboratory.)

1. *Overview of the Project*, Burren, J. W. (ed.)
2. *Universe Network Protocol Architecture*, Adams, C. J. (ed.)
3. *Network Monitoring, Management and Control*, Wilbur, S. R. (ed.)
4. *A Reconfigurable Fault-Tolerant Ring Network*, Kirk, P. R. (ed.)
5. *Remote File Access Protocol*, Pope, N. H. (ed.)
6. *The Provision of Authentication Facilities in Project Universe*, Pope, N. H. (ed.)
7. *Encryption Experiment*, Jackson, A. M., McEvoy, N. A., and Newman, B. B.
8. *The TRIPOS Filing Machine, a front end to a File Server*, Richardson, M. F. and Needham, R. M.
9. *Full Screen Access to an IBM Machine via Universe*, Adams, G. C.
10. *Remote Support for System Development and Operation*, Burren, J. W.
11. *Voice*, Adams, C. J. (ed.)
12. *Images*, Griffiths, J. W. R. (ed.)
13. *Teletex*, Kirstein, P. T. (ed.)
14. *Videotex*, Tolcher, D. J. (ed.)
15. *Universe Support Environment*, Hill, R. S. (ed.)
16. *File Distribution*, Cooper, C. S. (ed.)
17. *Satellite Bridge*, Waters, A. G. (ed.)
18. *Demonstrations*, Tuck, W. R. (ed.)
19. *Infrastructure*, Cooper, C. S. (ed.)
20. *Communications Development: Satellite Modem and Optical Fibre Link*, Griffiths, J. W. R. (ed.)
21. *The Project in Retrospect*, Burren, J. W. (ed.)

Appendix C. Participants in the Project

British Telecom Research Laboratories

S Alexander
S P Alvey
*G H L Childs**
J R E Cubbin
R Daniel
B Dew
K R G Fowler
L J Harvey
G D Mills
G Morrow
K W Rayner
A J T Richardson
P A Rout
M G Rowlands
A H Spittle
*D J Tolcher**
S M West
S J Wilders
C D Whitlum

Cambridge University

S Ades
C G Girling
I M Leslie
*R M Needham**
N J Ody
M F Richardson
D L Tennenhouse

GEC Marconi Research Centre

S R Forman
R J Godbold
M W Hall
A M Jackson
*D King**
P R Kirk
N J L Lockett
S P Lord
G Oddy
H J Pipe
N H Pope
D C Robinson
D S N Saul
D Shaughnessy
A N Slater
R Smith
S L Tavendale
I Vessey
P J Wilkinson
P G Wright

Logica Ltd

W Breach
R Butler
*P Campbell-Smith**
N Cartwright
D Cohen
D Dimmock
D Duncan
R Fine
J Forster
K Haines
R Hill
E Holland
A Hughes
D James
S James
R Jewsbury
J Kara
N Kennedy
W Lees
*C MacDowell**
R MacKay
N McEvoy
W Newman
R Perkins
J Pierce
P Radford
M Rich
S Shah
N Trier

Loughborough University of Technology

D S Chan
T N Chen
A P Clark
A F Erwood
M J Fairfield
W C Gabb
*J W R Griffiths**
B S Lee
F McVerry
D J Parish
P J Patrick
R Thirlby

Rutherford Appleton Laboratory

C J Adams
G C Adams
*J W Burren**
C S Cooper
*P F Linington***
A D Marshall
J S Robinson
J F Taylor
A R Thorne
A G Waters

University College London

H Armand
B S Bacarisse
R H Cole
C Campbell
T Daniel
H R Gamble
R J Hall
C J Kennington
*P T Kirstein**
G Knight
T M Liew
P J Lloyd
Z Ma
F Pintos
I Prince
B Segal
W Tuck
S R Wilbur
R P Winfield
P Williamson

* Principal Investigators
** Project Co-ordinator

(The list above includes all those who worked on the project at any time. Some worked full-time for the whole of the project; others were part-time or for part of the duration of the project.)

Steering Committee members

S L H Clarke
A J Egginton
L R H Harris
E B Holloway
R A Lingard
P F Linington
D W Mann
G Manning
*J H H Merriman**
G H Stafford
D B Thomas
W Wigglesworth
P R Williams
J C Williams
D J Withers

* Chairman

Appendix D. Project Bibliography

Adams, C. J. (1984). The UNIVERSE project. *Information Technology and the Computer Network,* (ed. K. G. Beauchamp), 113–23. Springer-Verlag, Berlin.

Adams, C. and Ades, S. (1985). Voice Experiments in the UNIVERSE Project. *Proceedings of International Conference on Communications 1985,* **2**, 927–35. IEEE, New York.

Adams, C. J., Adams, G. C., Waters, A. G., Leslie, I., and Kirk, P. (1982). Protocol Architecture of the UNIVERSE Project. *Pathways to the Information Society, Proceedings of ICCC 1982,* 379–83. North-Holland, Amsterdam.

Adams, G. C., Burren, J. W., Cooper, C. S., and Girard, P. M. (1982). The Interconnection of Local Area Networks via a Satellite Network. *New Advances in Distributed Computing Systems,* (ed. K. G. Beauchamp), 201–10. D. Reidel, Dordrecht.

Adams, C., Adams, G., Ades, S., Erwood, A., Fairfield, M., Griffiths, J., Patrick, P., and McEvoy, N. (1984). Voice and image communication in Project Universe. *Networks 84, Proceedings of European Computer Communications Conference,* 25–41. Online, London.

Burren, J. W. (1985). The use of satellite channels for computer networking. *Proceedings of International Symposium on Satellite Transmissions,* ESA SP-245, 71-8.

Burren, J. W. (1986). High-speed computer networking by satellite: a review of the results from Project Universe. *Journal of the IERE,* **56**, No. 5, 187-91.

Chen, T. N. (1985). Ring network reliability and a fault-tolerant Cambridge ring architecture. *Proc. NEOS85, Networks and Electronic Office Systems,* 23–30. IERE, London.

Cole, R. and Lloyd, P. (1983). A Flexible Architecture for Protocol Studies in a Multi-Network Environment. *Information Processing 83, Proceedings of IFIP 9th World Computer Congress,* (ed. R. E. A. Mason), 401–6. North-Holland, Amsterdam.

Cole, R., Adams, C., Celandroni, N., Ferro, E., Lenzini, L., and Leslie, I. (1983). An Interworking Architecture for STELLA and UNIVERSE.

Satellite and Computer Communications, Proceedings of IFIP-TC6 International Symposium, Versailles, 369–77. INRIA, Rocquencourt.

Cooper, C. S. (1984). Managed File Distribution on the UNIVERSE Network. *Communications Architectures & Protocols, SIGCOMM '84, ACM Computer Communication Review,* **14**, No. 2, 10–17.

Crowcroft, J. and Hall, R. (1984). A Distributed Approach to Loading in a Network Environment. *Ring Technology Local Area Networks, Proceedings of IFIP WG 6.4/University of Kent Workshop on Ring Technology based Local Area Networks,* (eds. I. N. Dallas and E. B. Spratt), 111–22. North-Holland, Amsterdam.

Daniel, R. (1983). The Universe Project. *Forum 83, 4th World Telecommunication Forum,* Part II, 2.7.10.1–4. ITU, Geneva.

Daniel, R. (1983). A review of Project Universe. *Proceedings of NEOS83, Networks and Electronic Office Systems,* 7–10. IERE, London.

Girling, C. G. (1982). Authentication in Project UNIVERSE. *Pathways to the Information Society, Proceedings of ICCC 1982,* 395–400. North-Holland, Amsterdam.

Griffiths, J. W. R. (1983). Voices and images experiments in Project Universe. *Proceedings of NEOS83, Networks and Electronic Office Systems,* 11–13. IERE, London.

Hill, R. (1984). The Universe support environment. *Networks 84, Proceedings of European Computer Communications Conference,* 15–24. Online, London.

Jackson, A. M., McEvoy, N. A., and Newman, B. B. (1984). Project UNIVERSE encryption experiment. *International Conference on Secure Communication Systems,* 14–19. IEE, London.

Kirstein, P. T. (1983). Some Experiences with Packet Switched Satellite Computer Networks. *EUTECO—European Teleinformatics Conference,* (ed. T. Kalin), 383–93. Elsevier Science Publishers, Amsterdam.

Kirstein, P. T. and Wilbur, S. R. (1983). Project Universe—lessons for the 1990s. *Data Communications: The Wired Society, State of the Art report,* **11**:8iii. Pergamon Infotech, Maidenhead.

Kirstein, P. T., Burren, J., Daniels, R., Griffiths, J. W. R., King, D., McDowell, C., and Needham, R. (1982). The UNIVERSE Project. *Pathways to the Information Society, Proceedings of ICCC 1982,* 442–7. North-Holland, Amsterdam.

Lees, W. (1983). The Universe network at Logica. *Local Networks—Strategy & Systems, Proceedings of Localnet '83 (Europe),* 311–26. Online, London.

Leslie, I. M. and Burren, J. W. (1985). Experiences with a satellite based, high capacity, packet switched network. *Computer Communications—AFRICOM '84, Proceedings of IFIP TC 6 1st African Conference on*

Computer Communications, (ed. F. Kamoun), 15–24. North-Holland, Amsterdam.

Leslie, I. M., Needham, R. M., Burren, J. W., and Adams, G. C. (1984). The Architecture of the UNIVERSE Network. *Communications Architectures & Protocols, SIGCOMM '84, ACM Computer Communication Review,* **14**, No. 2, 2–9.

Lloyd, P. J. and Cole, R. H. (1983). A comparative study of protocol performance on the UNIVERSE and SATNET satellite systems. *Satellite and Computer Communications, Proceedings of IFIP-TC6 International Symposium, Versailles,* 353–68. INRIA, Rocquencourt.

Newman, W. B. (1982). Design of an Encryption System for Project UNIVERSE. *Pathways to the Information Society, Proceedings of ICCC 1982,* 384–9. North-Holland, Amsterdam.

Richardson, M. F. and Needham, R. M. (1983). The TRIPOS Filing machine, a Front End to a File Server. *ACM Operating Systems Review,* **17**, No. 5, 120–7.

Rout, P. (1984). Business applications of Project Universe. *Networks 84, Proceedings of European Computer Communications Conference,* 1–13. Online, London.

Sorensen, S. (1983). Simulation Study of a UNIVERSE type Network. *Satellite and Computer Communications, Proceedings of IFIP-TC6 International Symposium, Versailles,* 343–52. INRIA, Rocquencourt.

Tuck, W., Richardson, T., Rich, M., and Prince, I. (1985). The Teletex Experiment on Project UNIVERSE. *The New World of the Information Society, Proceedings of ICCC 1984,* 359–63. North-Holland, Amsterdam.

Waters, A. G. (1983). Project Universe—implementation and experience of the satellite bridge. *Proceedings of NEOS83, Networks and Electronic Office Systems,* 15–20. IERE, London.

Waters, A. G. (1984). The Performance of the Satellite Bridge in the Universe Project. *Performance of Computer-Communication Systems, Proceedings of IFIP WG 7.3/TC 6 2nd International Symposium on the Performance of Computer-Communication Systems,* (eds. H. Rudin and W. Bux), 383–94. Elsevier Science Publishers, Amsterdam.

Waters, A. G. and Adams, C. J. (1984). The Satellite Transmission Protocol of the UNIVERSE Project. *Communications Architectures & Protocols, SIGCOMM '84, ACM Computer Communication Review,* **14**, No. 2, 18–24.

Waters, A. G., Adams, C. J., Leslie, I. M., and Needham, R. M. (1984). The use of Broadcast Techniques on the UNIVERSE Network. *Communi-*

cations Architectures & Protocols, SIGCOMM '84, ACM Computer Communication Review, **14**, No. 2, 52–7.

Wilbur, S. R. (1983). Initial experience with Universe at University College London. *Local Networks—Strategy & Systems, Proceedings of Localnet '83 (Europe),* 297–309. Online, London.

Wilbur, S. R. (1983). Program support for Project Universe. *Software & Microsystems,* **2**, No. 1, 3–8. IEE, London.

Wilbur, S. R. (1984). Aspects of Distributed System management. *Ring Technology Local Area Networks, Proceedings of IFIP WG 6.4/University of Kent Workshop on Ring Technology based Local Area Networks,* (eds. I. N. Dallas and E. B. Spratt), 123–32. North-Holland, Amsterdam.

Wilbur, S. R. and Ma, Z. (1982). An Outline of Network Measurements in Project UNIVERSE. *Pathways to the Information Society, Proceedings of ICCC 1982,* 390–4. North-Holland, Amsterdam.

Winfield, R. (1984). Document transfer by satellite. *Aslib Proceedings,* **36**, No. 4, 177–85.

References

Adams, C. and Ades, S. (1985). Voice Experiments in the UNIVERSE Project. *Proceedings of International Conference on Communications 1985,* **2**, 927–35. IEEE, New York.

Adams, G. C., Burren, J. W., Cooper, C. S., and Girard, P. M. (1982). The Interconnection of Local Area Networks via a Satellite Network. *New Advances in Distributed Computing Systems,* (ed. K. G. Beauchamp), 201–10. D. Reidel, Dordrecht.

Adams, J. L. and Falconer, R. M. (1984). 'Orwell': a protocol for carrying integrated services on a digital communications ring. *Electronics Letters,* **20**, No. 23, 970–71.

Andrews, D. W. and Schultz, G. D. (1982). A Token-Ring Architecture for Local-Area Networks: An Update. *Proceedings of Computer Networks, Compcon82,* 615–24. IEEE, New York.

Azouz, L., Huitema, C., and Kamoun., F. (1983). Performance evaluation of NADIR bulk data transmission protocol for a point to point satellite link. *IFIP–TC6 International Symposium on Satellite and Computer Communications,* 75–88. INRIA, Rocquencourt.

Azouz, L., Youssef, H., Huitema, C., and Kamoun, F. (1984). Performance evaluation of NADIR bulk data transmission protocol for a multipoint satellite link. *Performance of Computer-Communication Systems, Proceedings of IFIP WG 7.3/TC 6 2nd International Symposium on the Performance of Computer-Communication Systems,* (eds. H. Rudin and W. Bux), 367–82. Elsevier Science Publishers, Amsterdam.

Backes, F. (1988). Transparent Bridges for Interconnection of IEEE 802 LANS. *IEEE Network,* **2**, No. 1, 5–9.

Beker, H. and Piper, F. (1982). *Cipher Systems: The Protection of Communications.* Northwood, London.

Berntsen, J. A., Davin, J. R., Pitt, D. A., and Sullivan, N. G. (1985). MAC Layer Interconnection of IEEE 802 Local Area Networks. *Computer Networks and ISDN Systems,* **10**, No. 5, 259–73.

Birrell, A. D., Levin, R., Needham, R. M., and Schroeder, M. D. (1982). Grapevine: An Exercise in Distributed Computing. *Communications of the ACM,* **25**, No. 4, 260–74.

Böschen, R. (1982). PLANET—ein Private-Local-Area-Network. *Nachrichten Elektronik,* **36**, Heft 6, 255–6.

Bullington, K. and Fraser, J. M. (1959). Engineering Aspects of TASI. *The Bell System Technical Journal,* **38**, March, 353–64.

Burren, J. W. (1985). The use of satellite channels for computer networking. *Proceedings of International Symposium on Satellite Transmissions,* ESA SP-245, 71-8.

Burton, D. M. (1980). *Elementary number theory.* Allyn and Bacon, Newton, Massachusetts.

Bux, W. and Grillo, D. (1985). Flow Control in Local-Area Networks of Interconnected Token Rings. *IEEE Transactions on Communications* **COM-33**, No. 10, 1058–66.

Carrington, M. J. (1983). Local Area Networks—Meeting the users' needs. *Proceedings of NEOS83, Networks and Electronic Office Systems,* 151–5. IERE, London.

CCITT (1981*a*). Interface between data terminal equipment (DTE) and data circuit-terminating equipment (DCE) for terminals operating in the packet mode on public data networks. *Data communication networks; services and facilities, terminal equipment and interfaces,* Yellow Book, **VIII**, Fascicle VIII.2, Recommendation X.25, 100–190. ITU, Geneva.

CCITT (1981*b*). Procedures for the exchange of control information and user data between a packet assembly/disassembly facility (PAD) and a packet mode DTE or another PAD. *Data communication networks; services and facilities, terminal equipment and interfaces,* Yellow Book, **VIII**, Fascicle VIII.2, Recommendation X.29, 220–31. ITU, Geneva.

CCITT (1981*c*). Mean one-way propagation time. *General characteristics of international telephone connections and circuits,* Yellow Book, **III**, Fascicle III.1, Recommendation G.114, 70–2. ITU, Geneva.

CCITT (1981*d*). Teletex service. *Telegraph and telematic services operations and tarriffs,* Yellow Book, **II**, Fascicle II.4, Recommendation F.200, 281–300. ITU, Geneva.

CCITT (1981*e*). Teletex terminals. *Telegraph and telematic services terminal equipment,* Yellow Book, **VII**, Fascicle VII.2, Part I, Section 2, 59–164. ITU, Geneva.

CCITT (1987). Message Handling—System and Service Overview. *CCITT Draft Recommendation F.400/X.400, (X.mhs0),* Version 2 – May 1987. (Also referenced as ISO/TC97/SC18/WG4 N616.)

CCITT (1988). Framework for the provision of additional packet-mode bearer services. (Private communication; to appear as Recommendation I.122 in the Blue Book to be published by the ITU, Geneva.)

Celandroni, N., Ferro, E., Lenzini, L., Segal, B. M., and Olofsson, K. S. (1983). STELLA – Satellite Interconnection of Local Area Networks: Architecture and Protocols. *IFIP–TC6 International Symposium on Satellite and Computer Communications,* 25–40. INRIA, Rocquencourt.

Clark, D. D., Lambert, M. L., and Zhang, L. (1987). NETBLT: A High Throughput Transport Protocol. *Frontiers in Computer Communications Technology, SIGCOMM '87, ACM Computer Communication Review,* **17**, No. 5, 353–9.

Clarke, R. J. (1985). *Transform Coding of Images.* Academic Press, Orlando, Florida.

Cole, R., Adams, C., Celandroni, N., Ferro, E., Lenzini, L., and Leslie, I. (1983). An Interworking Architecture for STELLA and UNIVERSE. *IFIP–TC6 International Symposium on Satellite and Computer Communications,* 369–77. INRIA, Rocquencourt.

Comer, D. E. (1988). *Internetworking with TCP/IP: Principles, Protocols, and Architecture.* Prentice-Hall, Englewood Cliffs, New Jersey.

Cooper, C. S. (1984). Managed File Distribution on the UNIVERSE Network. *Communications Architectures & Protocols, SIGCOMM '84, ACM Computer Communication Review,* **14**, No. 2, 10–17.

Cooper, C. S. (1987). Design of a communications network subsystem for the Unidata project. *Journal of the IERE,* **57**, No. 3, 96–100.

Coulouris, G. F. and Dollimore, J. (1988). *Distributed Systems: Concepts and Design.* Addison-Wesley, Reading, Massachusetts.

Daka, J. S. J. and Waters, A. G. (1988). A High Performance Broadcast File Transfer Protocol. *Communications Architectures & Protocols, SIGCOMM '88, ACM Computer Communication Review,* **18**, No. 4, 274–81.

Dallas, I. N. (1981). *Transport Service Byte Stream Protocol (TSBSP).* UKC Computing Laboratory Report No. 1 (revision 3, 17th August 1981). University of Kent at Canterbury.

Davies, D. W. (1979). Cryptography and Crypto-Systems. *Interlinking of Computer Networks,* (ed. K. G. Beauchamp), 201–37. D. Reidel, Dordrecht.

Day, J. D. and Zimmermann, H. (1983). The OSI Reference Model. *Proceedings of the IEEE,* **71**, No. 12, 1334–40.

DCPU (1981*a*). *A Network Independent File Transfer Protocol.* Data Communication Protocols Unit of Department of Trade and Industry. (Available through the Joint Network Team, Rutherford Appleton Laboratory, UK.)

DCPU (1981*b*). *A Network Independent Job Transfer and Manipulation Protocol.* Data Communication Protocols Unit of Department of Trade and Industry. (Available through the Joint Network Team, Rutherford Appleton Laboratory, UK.)

Denning, D. E. R. (1982). *Cryptography and Data Security.* Addison-Wesley, Reading, Massachusetts.

Dennis, J. B. and Van Horn, E. C. (1966). Programming Semantics for Multiprogrammed Computations. *Communications of the ACM,* **9**, No. 3, 143–55.

Devlin, K. (1984). *Microchip Mathematics—number theory for computer users.* Shiva Publishing Ltd.

Dion, J. (1980). The Cambridge File Server. *ACM Operating Systems Review,* **14**, No. 4, 26–35.

Dion, J. (1981). *Reliable Storage in a Local Network.* University of Cambridge PhD thesis. (Available from Cambridge University Computer Laboratory as Technical Report No. 16.)

Edmond, W., Blumenthal, S., Echenique, A., Storch, S., Calderwood, T., and Rees, T. (1986). The Butterfly Satellite IMP for the Wideband Packet Satellite Network. *Communications Architectures & Protocols, SIGCOMM '86, ACM Computer Communication Review,* **16**, No. 3, 194–203.

Fabry, R. S. (1974). Capability-Based Addressing. *Communications of the ACM,* **17**, No. 7, 403–12.

Falconer, R. M. (1984). A Study of Techniques for Enhancing the Reliability of Ring Local Area Networks (LANs). *Ring Technology Local Area Networks, Proceedings of IFIP WG 6.4/University of Kent Workshop on Ring Technology based Local Area Networks,* (eds. I. N. Dallas and E. B. Spratt), 169–86. North-Holland, Amsterdam.

Feher, K. (1983). *Digital Communications - satellite/earth station engineering.* Prentice-Hall, Englewood Cliffs, New Jersey.

Fisher, D., Maynard, P., and Alexander, S. (1988). Broadband ISDN. *Computer Communications,* **11**, No. 4, 185–90.

Gardner, F. M. (1979). *Phaselock Techniques.* John Wiley and Sons, New York.

Girling, C. G. (1982). Object Representation on a Heterogeneous Network. *ACM Operating Systems Review,* **16**, No. 4, 49–59.

Girling, C. G. (1983). *Representation and Authentication on Computer Networks.* University of Cambridge PhD thesis. (Available from Cambridge University Computer Laboratory as Technical Report No. 37.)

Gopal, I. S. and Jaffe, J. M. (1984). Point-to-Multipoint Communication Over Broadcast Links. *IEEE Transactions on Communications,* **COM-32**, No. 9, 1034–44.

Hamilton, K. G. (1984). *A Remote Procedure Call System*. University of Cambridge PhD thesis. (Available from Cambridge University Computer Laboratory as Technical Report No. 70.)

Hamming, R. W. (1980). *Coding and information theory*. Prentice-Hall, Englewood Cliffs, New Jersey.

Hardy, G. H. and Wright, E. M. (1979). *An introduction to the theory of numbers*. Oxford University Press.

Hart, J. (1988). Extending the IEEE 802.1 MAC Bridge Standard to Remote Bridges. *IEEE Network*, **2**, No. 1, 10–15.

Hart, J. H. (1985). 'Bridges' smooth troubled waters for wide-area networking. *Data Communications*, **14**, No. 3, 209–20.

Hawe, B., Kirby, A., and Stewart, B. (1984). Transparent Interconnection of Local Area Networks with Bridges. *Journal of Telecommunication Networks*, **3**, No. 2, 116–30.

Heggestad, H. M. (1984). Voice and Data Communications on an Experimental Wideband Internetwork System. *Journal of Telecommunication Networks*, **3**, No. 2, 131–46.

Heggestad, H. M. and Weinstein, C. J. (1983). Voice and data communication experiments on a wideband satellite/terrestrial internetwork system. *Proceedings of International Conference on Communications 1983*, **1**, 1–8. IEEE, New York.

Hine, M. G. N. (1984). *The STELLA Experiment Final Report*. Commission of European Communities, EUR 9090.

Hine, M., Adams, C., Burren, J., Celandroni, N., Ferro, E., Koudelka, O., Riedler, W., and Waibel, T. (1988). SATINE-2 — an Experiment in High Speed Wide Area Networking. *Research into Networks and Distributed Applications, EUTECO '88*, (ed. R. Speth), 577–605. North-Holland, Amsterdam.

Hopper, A., Temple, S., and Williamson, R. C. (1986). *Local Area Network Design*. Addison-Wesley, Reading, Massachusetts.

Huitema, C. and Radureau, J. (1982). Interconnecting Local Area Networks through Satellite Links. *Pathways to the Information Society, Proceedings of ICCC 1982*, 431–5. North-Holland, Amsterdam.

IBM (1979). *IBM Virtual Machine Facility/370: System Programmer's Guide*, GC20-1807-7. IBM.

IBM (1980). *IBM 3270 Information Display System: 3271 Control Unit, 3272 Control Unit, 3275 Display Station Description and Programmer's Guide*, GA23-0060-0. IBM.

IBM (1987). *IBM 3270 Information Display System: Introduction*, GA27-2739-22. IBM.

IEEE (1984). *IEEE Standards for Local Area Networks: Logical Link Control.* IEEE and John Wiley and Sons, New York.

IEEE (1985*a*). *IEEE Standards for Local Area Networks: Carrier Sense Multiple Access with Collision Detection (CSMA/CD) Access Method and Physical Layer Specifications.* IEEE and John Wiley and Sons, New York.

IEEE (1985*b*). *IEEE Standards for Local Area Networks: Token-Passing Bus Access Method and Physical Layer Specifications.* IEEE and John Wiley and Sons, New York.

IEEE (1985*c*). *IEEE Standards for Local Area Networks: Token Ring Access Method and Physical Layer Specifications.* IEEE and John Wiley and Sons, New York.

ISO (1983). *Information Processing Systems – Open Systems Interconnection – Basic Reference Model.* ISO 7498.

ISO (1985*a*). *Information Processing Systems – Open Systems Interconnection – Job Transfer and Manipulation concepts and services.* ISO/DIS 8831.

ISO (1985*b*). *Information Processing Systems – Open Systems Interconnection – Specification of the basic class protocol for Job Transfer and Manipulation.* ISO/DIS 8832.

ISO (1986). *Internal Organization of the Network Layer.* ISO/DIS 8648. (Private communication of a working draft of this standard.)

ISO (1987). *Information Processing Systems – Open Systems Interconnection – File Transfer, Access, and Management.* ISO 8571.

ISO (1988). *Local Area Networks – Part 7: Slotted Ring Access Method and Physical Layer Specification.* ISO 8802-7. (Private communication of the final text of this standard.)

Jacobs, I. M., Binder, R., and Hoversten, E. V. (1978). General Purpose Packet Satellite Networks. *Proceedings of the IEEE,* **66**, No. 11, 1448–67.

Jacobs, I. M., Hoversten, E. V., Abel, N. L., Binder, R., Bressler, R. D., Edmond, W. B., and Killian, E. A. (1979). Packet Satellite Network Design Issues. *Conference Record, National Telecommunications Conference, Washington,* **3**, 45.2.1–12. IEEE, New York.

Jain, A. (1981). Image Data Compression: A Review. *Proceedings of the IEEE,* **69**, No. 3, 349–89.

Jayant, N. S. and Noll, P. (1984). *Digital Coding of Waveforms.* Prentice-Hall, Englewood Cliffs, New Jersey.

Jennings, D. M., Landweber, L. H., Fuchs, I. H., Farber, D. J., and Adrion, W. R. (1986). Computer Networking for Scientists. *Science,* **231**, 943–50.

Kahn, R. E. (1979). The Introduction of Packet Satellite Communications. *Conference Record, National Telecommunications Conference, Washington,* **3**, 45.1.1–8. IEEE, New York.

Kuo, F. F. (*ed.*) (1981). *Protocols and Techniques for Data Communication Networks.* Prentice-Hall, Englewood Cliffs, New Jersey.

Larmouth, J. (*ed.*) (1982). *Cambridge Ring 82 – protocol specifications.* UK Joint Network Team.

Leslie, I. M. (1983). *Extending the Local Area Network.* University of Cambridge PhD thesis. (Available from Cambridge University Computer Laboratory as Technical Report No. 43.)

Leslie, I. M., Needham, R. M., Burren, J. W., and Adams, G. C. (1984). The Architecture of the UNIVERSE Network. *Communications Architectures & Protocols, SIGCOMM '84, ACM Computer Communication Review,* **14**, No. 2, 2–9.

Linington, P. F. (1983). Fundamentals of the Layer Service Definitions and Protocol Specifications. *Proceedings of the IEEE,* **71**, No. 12, 1341–5.

Littlewood, M., Gallagher, I. D., and Adams, J. L. (1987). Evolution toward an ATD multi-service network. *British Telecom Technology Journal,* **5**, No. 2, 52–62.

Meunier, P., Jourdan, F., Minet, J. F., and Valet, I. (1983). An experimental bulk transport service using a satellite link in DSA environment. *IFIP–TC6 International Symposium on Satellite and Computer Communications,* 43–50. INRIA, Rocquencourt.

NBS (1977). *Specifications for the Data Encryption Standard.* Federal Information Processing Standards Publication 46. US National Bureau of Standards.

Needham, R. M. and Herbert, A. J. (1982). *The Cambridge Distributed Computing System.* Addison-Wesley, Reading, Massachusetts.

Needham, R. M. and Walker, R. D. H. (1977). The Cambridge CAP Computer and its protection system. *Operating Systems Review,* **11**, No. 5, 1–10.

Nelson, B. J. (1981). *Remote Procedure Call.* Carnegie Mellon University PhD thesis; CMU Report CMU-CS-81-119.

Nessett, D. M. (1983). A Systematic Methodology for Analyzing Security Threats to Interprocess Communication in a Distributed System. *IEEE Transactions on Communications,* **COM-31**, No. 9, 1055–63.

Netravali, A. N. and Limb, J. O. (1980). Picture coding: A Review. *Proceedings of the IEEE,* **68**, No. 3, 366–406.

Newman, R. M., Budrikis, Z. L., and Hullett, J. L. (1988). The QPSX Man. *IEEE Communications Magazine,* **26**, No. 4, 20–28.

Owen, F. F. E. (1982). *PCM and Digital Transmission Systems.* McGraw-Hill, New York.

Pope, N. H., Tolcher, D. J., Wilbur, S. R., and Rosner, R. A. (1986). Project ADMIRAL: research into networks & distributed systems. *Networks 86,* 301–9. Online, London.

Pierce, J. R. and Posner, E. C. (1980). *Introduction to Communication Science and Systems.* Plenum Press, New York.

PSS User Forum (1980). *A Network Independent Transport Service.* British Telecom. (Available through the Joint Network Team, Rutherford Appleton Laboratory, UK.)

PSS User Forum (1981). *Character Terminal Protocols on PSS.* British Telecom. (Available through the Joint Network Team, Rutherford Appleton Laboratory, UK.)

Rabin, M. O. (1976). Probabilistic Algorithms. *Algorithms and Complexity: new directions and recent results,* (ed. J. F. Traub), 21–39. Academic Press, Orlando, Florida.

Rice, R. F. (1975). An advanced imaging communication system for planetary exploration. *Efficient Transmission of Pictorial Information, SPIE Seminar Proceedings,* **66**, 70–89.

Rice, R. F. and Plaunt, J. R. (1981). Adaptive Variable-Length Coding for Efficient Compression of Spacecraft Television Data. *IEEE Transactions on Communications,* **COM-19**, No. 6, 889–97.

Richards, M. and Whitby-Strevens, C. (1979). *BCPL – the language and its compiler.* Cambridge University Press.

Richards, M., Aylward, A. R., Bond, P., Evans, R. D., and Knight, B. J. (1979). TRIPOS—A Portable Operating System for Mini-computers. *Software—Practice and Experience,* **9**, No. 7, 513–26.

Richardson, M. F. and Needham, R. M. (1983). The TRIPOS Filing Machine, a front end to a File Server. *ACM Operating Systems Review,* **17**, No. 5, 120–7.

Rivest, R. L., Shamir, A., and Adleman, L. (1978). A Method for Obtaining Digital Signatures and Public-Key Cryptosystems. *Communications of the ACM,* **21**, No. 2, 120–6.

Ross, F. E. (1986). FDDI—a Tutorial. *IEEE Communications Magazine,* **24**, No. 5, 10–17.

Rutkowski, A. M. (1985). *Integrated Services Digital Networks.* Artech House, Dedham, Massachusetts.

Sabnani, K. and Schwartz, M. (1985). Multidestination Protocols for Satellite Broadcast Channels. *IEEE Transactions on Communications,* **COM-33**, No. 3, 232–40.

Saltzer, J. H. (1982). On the Naming and Binding of Network Destinations. *Local Computer Networks, Proceedings of IFIP TC6 International Symposium on Local Computer Networks,* (eds. P. Ravasio, G. Hopkins, and N. Naffah), 311–17. North-Holland, Amsterdam.

Saltzer, J. H. and Pogran, K. T. (1980). A Star-Shaped Ring Network with High Maintainability. *Computer Networks,* **4**, No. 5, 239–44.

Saltzer, J. H., Clark, D. D., and Pogran, K. T. (1981). Why a Ring? *Proceedings of Seventh Data Communications Symposium,* 211-17. IEEE, New York.

Schroeder, M. D., Birrell, A. D., and Needham, R. M. (1984). Experience with Grapevine: The Growth of a Distributed System. *ACM Transactions on Computer Systems,* **2**, No. 1, 3–23.

Sharpe, W. P. and Cash, A. R. (*eds.*) (1982). *Cambridge Ring 82 – interface specifications.* UK Science and Engineering Research Council and Joint Network Team.

Shepherd, W. D. (1981). Ancilla—A Server for the Cambridge Model Distributed System. *Software—Practice and Experience,* **11**, No. 11, 1185–95.

Shoch, J. F. (1978). Inter-Network Naming, Addressing, and Routing. *Proceedings of Compcon78,* 72–9. IEEE, New York.

Shoch, J. F. and Hupp, J. A. (1980). Measured Performance of an Ethernet Local Network. *Communications of the ACM,* **23**, No. 12, 711–21.

Sinkoskie, W. D. and Cotton, C. J. (1988). Extended Bridge Algorithms for Large Networks. *IEEE Networks,* **2**, No. 1, 16–24.

SRI (1982*a*). *Internet Protocol Transition Workbook.* SRI International, Menlo Park, California.

SRI (1982*b*). *Internet Protocol Implementation Guide.* SRI International, Menlo Park, California.

Stallings, W. (1987). Local network standards. *Handbook of Computer-Communications Standards, Volume 2.* Howard W. Sams/Macmillan, Indianapolis, Indiana.

Symicron (1980). *The DTSX User Manual.* Symicron Ltd, London.

Tanenbaum, A. S. (1981). *Computer Networks.* Prentice-Hall, Englewood Cliffs, New Jersey.

Tanenbaum, A. S. (1988). *Computer Networks (Second Edition).* Prentice-Hall, Englewood Cliffs, New Jersey.

Tennenhouse, D. L., Leslie, I. M., Needham, R. M., Burren, J. W., Adams, C. J., and Cooper, C. S. (1987). Exploiting Wideband ISDN: The Unison Exchange. *Proceedings of the IEEE INFOCOM '87,* 1018–26. Computer Society of the IEEE, Washington.

Thompson, T. and Baran, N. (1988). The NeXT Computer. *Byte,* **13**, No. 12, 158–75.

Valet, I. (1983). Bulk transfer protocols on satellite link: study within the OSI reference model. *IFIP–TC6 International Symposium on Satellite and Computer Communications,* 65–74. INRIA, Rocquencourt.

Viterbi, A. J. and Omura, J. K. (1979). *Principles of Digital Communication and Coding.* McGraw-Hill, New York.

Voydock, V. L. and Kent, S. T. (1983). Security Mechanisms in High-Level Network Protocols. *ACM Computing Surveys,* **15**, No. 2, 135–71.

Waters, A. G. (1984). The Performance of the Satellite Bridge in the Universe Project. *Performance of Computer-Communication Systems, Proceedings of IFIP WG 7.3/TC 6 2nd International Symposium on the Performance of Computer-Communication Systems,* (eds. H. Rudin and W. Bux), 383–94. Elsevier Science Publishers, Amsterdam.

Waters, A. G. and Adams, C. J. (1984). The Satellite Transmission Protocol of the UNIVERSE Project. *Communications Architectures & Protocols, SIGCOMM '84, ACM Computer Communication Review,* **14**, No. 2, 18–24.

Waters, A. G., Adams, C. J., Leslie, I. M., and Needham, R. M. (1984). The use of Broadcast Techniques on the UNIVERSE Network. *Communications Architectures & Protocols, SIGCOMM '84, ACM Computer Communication Review,* **14**, No. 2, 52–7.

Wells, M. (1984). The JANET Project. *University Computing,* **6**, 56–62.

Wilkes, M. V. and Needham, R. M. (1979). *The Cambridge CAP Computer and Its Operating System.* Elsevier North Holland, Amsterdam.

Wilkes, M. V. and Wheeler, D. J. (1979). The Cambridge digital communication ring. *Proceedings of Local Area Communications Network Symposium,* 47–60. Mitre Corporation and National Bureau of Standards, Boston.

Zafiropulo, P. (1974). Performance Evaluation of Reliability Improvement Techniques for Single-Loop Communications Systems. *IEEE Transactions on Communications,* **COM-22**, No. 6, 742–51.

Index